# Environmental Forensics
## Proceedings of the 2014 INEF Conference

Edited by

**Gwen O'Sullivan**
*Mount Royal University, Calgary, Canada*
*Email: gosullivan@mtroyal.ca*

**David Megson**
*Ryerson University, Toronto, Canada*
*Email: David.Megson@ontario.ca*

The proceedings of the INEF Cambridge Conference 2014 held at St John's College, Cambridge, UK on 4–6 August, 2014.

Special Publication No. 350

Print ISBN: 978-1-78262-444-8
PDF eISBN: 978-1-78262-507-0

A catalogue record for this book is available from the British Library

© The Royal Society of Chemistry 2015

*All rights reserved*

*Apart from any fair dealing for the purpose of research or private study for non-commercial purposes, or criticism or review as permitted under the terms of the UK Copyright, Designs and Patents Act, 1988 and the Copyright and Related Rights Regulations 2003, this publication may not be reproduced, stored or transmitted, in any form or by any means, without the prior permission in writing of The Royal Society of Chemistry or the copyright owner, or in the case of reprographic reproduction only in accordance with the terms of the licences issued by the Copyright Licensing Agency in the UK, or in accordance with the terms of the licences issued by the appropriate Reproduction Rights Organization outside the UK. Enquiries concerning reproduction outside the terms stated here should be sent to The Royal Society of Chemistry at the address printed on this page.*

The RSC is not responsible for individual opinions expressed in this work.

Published by The Royal Society of Chemistry,
Thomas Graham House, Science Park, Milton Road,
Cambridge CB4 0WF, UK

Registered Charity Number 207890

Visit our website at www.rsc.org/books

Printed in the United Kingdom by CPI Group (UK) Ltd, Croydon, CR0 4YY, UK

# Foreword

Stephen Mudge

The world of Environmental Forensics is full of interesting people, real-world cases and techniques and many of these came together at Cambridge in August 2014 to showcase the state of the art. INEF was formed to provide a platform for practitioners to get together, physically and virtually, to advance our specialties. We had a meeting in St. John's College, Cambridge in 2011 and this was very successful both scientifically and socially hence our desire to go back to this location. I am pleased to say that both the location and weather delivered for us again and we had a great meeting. We had 85 participants from more than 15 countries with 45 presentations and workshop sessions; we are grateful to all those who made the journey to Cambridge to share their experiences and knowledge.

INEF is a charitable organisation under the Royal Society of Chemistry umbrella and we always endeavour to keep the costs as low as possible to cover the expense to running the event. Additionally, we try and support younger members of our community with reduced rates, where possible. We are exceptionally grateful to the companies that have sponsored us and enabled us to provide these lower rates and the activities at the meetings. In 2014, we wish to acknowledge; Restek, Markes international, Shimadzu, Thermo, JSB and RSCfor their new and continued support for us – please continue to view their websites and use their services when appropriate.

As practitioners of Environmental Forensics, we have an impressive armoury of tools available to us to do everything from identification of the sources, apportion responsibility, age date releases, determine "damage" and answer any manner of related questions. I find when talking to people new to our work, they are always amazed at what is now possible and your work, presentations and publications are a great way of getting this over to the outside community. It was not too long ago that we were unable to identify what was under the UCM hump in a GC trace – not only can we identify what is there now, we can use this to tell us a whole lot more about the processes in the environment. These and other great innovations are what make EF such an exciting area to work.

I would like to that all those who organised and facilitated the meeting and invite you to join us for our next meeting in Toronto, 2015.

# Author Biographies

**Gwen O'Sullivan** Dr Gwen O'Sullivan is an Assistant Professor of Environmental Science at Mount Royal University. Dr O'Sullivan earned a B.Sc. in Environmental Science from the University of Limerick and a PhD in Environmental Chemistry from Queen's University of Belfast. Over the course of her career, in industry, consultancy and academia, Dr O'Sullivan has developed technical expertise in the areas of environmental chemistry, environmental forensics, air quality and contaminated land and groundwater. She has worked on numerous research and industrial projects including the development of technologies and remedial actions plans for the treatment of petroleum hydrocarbons, chlorinated solvents, and  saline impacted sites. She has also designed and managed environmental forensic investigations involving compounds of concerns including drilling fluids, petroleum hydrocarbons, polycyclic aromatic hydrocarbons, polychlorinated biphenlys, polychlorinated dibenso-$p$-dioxins and dibenzofurans, methane and nitrates. She has also authored numerous scientific articles, edited books series and successfully competed for research grant both nationally and internationally.

**David Megson** Dr David Megson is a post-doctoral research fellow at Ryerson University who is conducting research at the Ontario Ministry of Environment and Climate Change. Dr Megson is currently investigating sources of legacy and emerging persistent organic pollutants and monitoring their fate and transport in the terrestrial and marine environment. He is undertaking the extraction of soils, biosolids, and marine mammal tissues and performing analysis using advanced mass spectrometry techniques including GCxGC-HRToFMS and FTICR. Dr Megson obtained a PhD at Plymouth University (UK), using multidimensional chromatography to create high resolution PCB signatures for  use as an ecological monitoring tool and to identify and age date human exposure to PCBs. David also holds a BSc in Environmental Forensics and an MSc in Environmental Analysis and Assessment. He has over four years experience working in the UK as an environmental consultant, specialising in human health risk assessment and contaminant fate and transport. He has undertaken contaminated land assessments, remediation projects and forensic investigations involving source identification of ground gas, PAHs, PCBs and PCDD/Fs. He has published several scientific papers and co-authored book chapters in the field of environmental forensics focusing on POPs in the environment. Dr Megson is a member of the Royal Society of Chemistry, the Society of Brownfield Risk Assessment and is an associate fellow of the higher education academy.

**Anne Brosnan** Anne Brosnan is the Deputy Director (Chief Prosecutor) Legal Services, Environment Agency, England. She is a qualified solicitor with Higher Rights of Audience in Criminal Proceedings and a Masters degree in Environmental Quality Management. In 2005 she spent a year in Sydney Australia on a working exchange with the then Department of Environment and Conservation. Anne established the Agency's Serious Casework Group in 2007 and was closely involved with the introduction the English environmental civil sanctions regime in 2010. Anne is Co President of the European Network of Prosecutors for the Environment (ENPE) founded in September 2012.

**Cristina Barbieri** Cristina Barazzetti Barbieri is a Criminalist at the Criminalistics Department/IGP, official forensic agency of Rio Grande do Sul State Government, since 1997. She started to work with environmental crimes investigation from the beginning of her career as a forensic expert. Cristina is a biologist and completed her undergraduate studies at PUC-RS, Porto Alegre, RS, Brazil. She has a Specialization Course in Environmental Quality Management by the same University and a MSc degree in Ecology at UFRGS. She is now finishing her  PhD in Nuclear Technology-Materials at IPEN/USP, where she develops a work related to environmental crimes forensics tracking sources of contaminants in sediments, using metals, isotopes and organics.

**Hsin-Lan Hsu** Dr Hsin-Lan Hsu is a senior researcher in the Department of Environmental Forensics at Industrial Technology and Research Institute, a leading research organization in Taiwan devoted to develop technologies for industries and the government. She is currently developing the compound-specific isotope analysis for carbon, hydrogen and chloride isotopes of organic compounds in groundwater, air and sediments and working closely with the Taiwan EPA and local consulting companies to identify the sources of chlorinated organic compounds in the groundwater. Dr Hsu obtained a PhD in the Environmental and Water Resources Engineering program at the University of Michigan, Ann Arbor, MI, where she studied the impact of surface-active impurities of organic liquid waste on the liquid waste transport in the groundwater. She also holds a BSc in Physics and MSc in Environmental Engineering at the National Taiwan University. Dr Hsu is specialized in chlorinated DNAPL contaminations and had spent more than seven years of career life undertaking site investigation, remediation, and forensic projects. She is a member of Taiwan Association of Soil and Groundwater Environmental Protection and Chinese Environmental Analytical Society.

**Jean Christophe Balouet** Dr Jean-Christophe Balouet is a former research scientist at the Paris Museum of Natural History and Smithsonian Institution. Over the past 26 years, he has served dozens of governmental, intergovernmental and non-governmental organizations and over 120 forensic cases worldwide. His expertise deals with air, soil and groundwater pollution. His interest in age-dating pollution has led him to develop  dendrochemical methods over the past 15 years. He is the author of over 100 peer-reviewed scientific articles. He serves environmental forensic organisations, such as ISEF as associate editor, or Compagnie Nationale des Experts Judiciaires en Environnement as Secretary General.

 **Robert Morrison** Dr Robert D. Morrison has a B.S. in Geology, a MS in Environmental Studies, a MS in Environmental Engineering, and a PhD in Soil Physics from the University of Wisconsin at Madison. Dr Morrison has worked for 45 years as an environmental consultant on projects related to soil and groundwater contamination, including site investigations and remediation. Dr Morrison currently specializes in the forensic review and interpretation of scientific data to identify the source and age of a contaminant release. Dr Morrison is the co-founder of the International Society of Environmental Forensics (ISEF) and the International Network of Environmental Forensics (INEF) and has served as Chief Editor of the *Journal of Environmental Forensics* and on the editorial board of *Ground Water* and *Ground Water Monitoring Review & Remediation*, among others. Dr Morrison wrote the first book on environmental forensics in 1999 and is credited with coining the term *environmental forensics* in the peer reviewed literature. Dr Morrison is the author or co-author of 15 books and numerous peer-reviewed articles, the majority of which address environmental forensic subjects

**Daniel Bouchard** Dr Daniel Bouchard obtained a MSc in Soil Science (Laval University, Canada) and a PhD in Hydrogeology (University of Neuchâtel, Switzerland). Daniel is currently a researcher at the Centre for Hydrogeology and Geothermic (University of Neuchâtel, Switzerland). He has over 10 years of experience specializing in compound-specific isotope analysis (CSIA). Over the last decade, he has contributed to expand CSIA application to gas-phase contaminant characterization present in the soil gas, to performance assessment of groundwater remediation treatments, and also developed new analytical methods combined with field sampling methods to collect gas-phase contaminants in view to apply CSIA in forensics studies.

 **John Bruce** John Bruce is currently working as a Knowledge Transfer Partnership Associate for the University of Portsmouth and DustScan Ltd, investigating methods of dust dispersion modelling for implementation into DustScan's business. His role includes research into empirical modelling of dust based on observed measurements and utilising air quality modelling software for use in dust predictions. He obtained a BSc in Environmental Forensics from the University of Portsmouth in 2012 and now has over two years' experience in environmental consultancy. Mr Bruce is an Associate Member of the Institute of Air Quality Management and the Institute of Environmental Sciences. He is currently studying towards a PhD at the University of Portsmouth with his research focusing on expanding and improving on the empirical aspects of his dust modelling for the KTP project.

**Rory Doherty** Rory is an environmental engineer / geoscientist whose interdisciplinary research and teaching is based on the role, fate, monitoring and management of contaminants. His research encompasses geogenic and anthropogenic sources and sinks of contamination as well as the emerging discipline of biogeophysics where geophysical techniques can be used to monitor subsurface microbial activity in large scale bioelectrical systems. Rory has extensive remediation experience with recent academic work in the investigation, monitoring and management of brownfield sites. At Queen's University Belfast he has worked on projects that have developed biological Permeable Reactive Barriers and ex situ chemox remediation approaches.

## Author Biographies

**Michael Wade** Dr Michael J. Wade is Principal Scientist of Wade Research, Inc.™, a small business that provides geochemical consulting services to a variety of government agencies, industrial clients, and law firms. Dr Wade is an organic geochemist with over 36 years post-doctoral experience with an overall total of 43 years of strong technical and project management experience in a variety of research programs with special emphasis on study of organic contamination in the environment. He regularly provides expert forensic services both through deposition process as well as testimony in various U.S. Federal and State Courts in the areas of environmental contamination, including assessment of sources of contamination, identification of petroleum product types, quantification of weathering effects on petroleum products, and age-dating of petroleum product releases. Since 1992, working through Wade Research, Inc., Dr Wade has engaged in the conduct of numerous projects dealing with the various aspects of environmental assessment, including measuring degradation of petroleum hydrocarbons and development of quantitative hydrocarbon fingerprinting techniques that identify sources of subsurface petroleum contamination. Recently, Dr Wade has devoted an increasing amount of time and effort to increasing the quantitative understanding of petroleum product weathering reactions in the environment. As part of his assignment mix, he has completed numerous assignment that have refined quantitative field and laboratory investigation approaches designed to establish time frames for the release of gasoline, kerosene, diesel fuel and heavier fuel oils in subsurface petroleum contamination cases. Annually, through Wade Research, Inc., Dr Wade conducts 20 to 30 such programs for clients throughout North America. In addition to his regular assignment mix, Dr Wade teaches forensic geochemical continuing education courses for a variety of state and professional society venues throughout the United States. Such courses provide today's environmental professionals with a broad background in organic chemistry that is then focussed down to specific tools that investigators can use to develop information on anthropogenic contaminants that lead to allocation discussions and/or legal responsibility resolution.

**Andrea D'Anna** Andrea D'Anna has 13 years of experience in Site Remediation and Engineering design; he holds a Masters degree in Environmental Science and has a Chemistry Expert certificate. Andrea joined AECOM in 2002 and joined the Process Engineering Group in 2006, and in 2009 began working on larger plants using innovative remediation technologies. During his career, Andrea has on worked on numerous remediation projects. His work has included preparation of risk analyses, coordination of design tasks, installation and O&M of multiple remediation systems such as SVE, AS, P&T, DPE, and TPE. Andrea has performed

LNAPL studies (environmental forensics – origin, timing, etc), prepared SCM for remediation technologies, sized remediation plants (biological, chemical-physic system, Activated Carbon), performed plant revamping and optimization, and prepared technical specifications. From two years Andrea worked on an innovative remediation systems including Oxygen Micro-bubbles injection, ERD system and In situ Thermal treatment). Principal Andrea's skills are: Theoretical knowledge of the matter, continuously updated (training), innovation technologies, wide range of remediation technologies; great experience in environmental sector, close relationship and trust with Clients/Supplier, negotiation skills with the Control Agencies, recognition by stakeholders (Control Agencies, local communities, universities, research institutions, associations, etc..) and lot of imagination (Think out of the Box). Andrea is a dynamic and collaborative person with a natural predisposition to give a hand to the next and team working and he is enthusiastic about innovative and sustainable Technologies.

**Funmilayo Doherty** Dr (Mrs) Doherty V. Funmilayo is an Environmental Toxicologist and a senior lecturer in the Environmental Biology unit, Yaba College of Technology. She attended Ahmadu Bello University, Zaria Nigeria, for her Bachelor's degree in Zoology and graduated with a second class upper. She obtained an MSc and PhD in Environmental Toxicology and Pollution Management from the University of Lagos, Nigeria. Her area of research interests include heavy metals monitoring and biological effects, impact of petroleum products spillage on aquatic and terrestrial ecosystems, biological effects of Hydrocarbons (BTEX) and identification of biomarkers of exposure to pollutants. Dr (Mrs) Doherty is a member of several professional bodies. She is a recipient of many awards, Scholarships and research grants, these include Federal Government Scholarship for Post graduate students in 2003, World bank Unilag Step-B Innovators of Tomorrow research grant in 2011, just to mention a few. She has over 30 publications in reputable national and international Journals. She has a passion for youths acquiring skills to develop themselves. Presently, Dr Doherty is the President, Society for Environmental Toxicology and Pollution Mitigation (SETPOM) and the Coordinator, Yaba College of Technology UNESCO-UNEVOC Centre for Research and Sustainable Development.

**M. K. Ladipo** Dr M. K. Ladipo is an analytical chemist and a chief lecturer in the Department of Polymer and Textile Technology, Yaba College of Technology with over 30 years experience. She attended Ahmadu Bello University, Zaria Nigeria, for her Bachelor's degree in Chemistry. She obtained an MSc in Polymer Science and Technology and PhD in analytical chemistry from the same University. Dr Ladipo is a member of several professional bodies including the Society for Environmental Toxicology and Pollution Mitigation (SETPOM), Polymer Institute of Nigeria

and Textile Society of Nigeria. She has several publications in reputable national and international Journals and has authored two books. She is a recipient of many awards; these include women in merit Gold award and Business Reports International Merit Award for Excellence in service to Humanity to mention a few. Her research interests include environmental monitoring of aquatic and terrestrial and effect of textile effluents on organisms and the environment.

**Stephanie Turnbull** Stephanie Marie Turnbull is a postgraduate researcher at the University of Glasgow who has recently started her PhD in the school of engineering. Stephanie's research focuses on the investigation of position specific isotope analysis which looks at the intramolecular structure of a molecule to determine their fate and transit during biodegradation. The aim is to further develop the methodology for the on-line system of analysis using a GC-c-IRMS. Stephanie obtained a first class honours BSc in Forensic Science from the University of Wolverhampton (UK) in 2014. Stephanie concentrated her interests specifically into environmental forensics as a result of her sponsored attendance at an INEF conference in 2011. INEF 2011 allowed Stephanie to effectively network in order to secure numerous international/national internships with large multinational companies like AkzoNobel. As a result of this industrial experience Stephanie wrote her undergraduate dissertation on an environmental investigation project. This project was further selected to be presented at the University of Coventry and Cambridge University as part of INEF 2014. Stephanie is a member of the Royal Society of Chemistry and the British Mass Spectrometry Society.

**Leo M. Rebele** Mr. Rebele is a vice president and senior environmental scientist with the Irvine office of Tetra Tech, Inc. With over 18 years of experience in the environmental industry, he possesses a broad background in environmental science. He has worked in collaboration with numerous municipal and private clients to apply cutting edge technical, programmatic and funding approaches to environmental issues on redevelopment projects. Mr. Rebele advocates for the use of environmental forensics as a practical means to solve issues as part of property investigations and transactions rather

than only using environmental forensics to support litigation projects. His experience includes the application of forensic techniques in establishing cleanup cost liability for commingled groundwater plumes of gasoline and chlorinated volatile organic compounds. He has also worked with waterfront property owners to establish potential sources of polynuclear aromatic hydrocarbons in soil and sediments. Mr. Rebele holds a Master of Science degree in Marine Resource Management from Oregon State University and undergraduate degree from the University of British Columbia in Biological Oceanography with an emphasis on nearshore pollution ecology.

# Thanks to our Conference Sponsors!

For over 25 years, Restek has been a leader in the development and manufacturing of GC and LC columns, reference standards, sample preparation materials, accessories, and more. Our reputation for unbeatable Plus 1 customer service and top-quality products is well known throughout the international chromatography community, and we are proud to provide analysts around the world with products and services to monitor the quality and safety of air, water, soil, food, pharmaceuticals, chemicals, and petroleum.

Markes International is a specialist provider of technologies and expertise that enable chemists to meet analytical challenges in the sampling and detection of trace-level organic compounds. As a long-standing leader in analytical thermal desorption, Markes manufactures a comprehensive range of instrumentation, sampling equipment and accessories that enhance the capability of GC–MS. Markes has also gained recognition for its BenchTOF range of time-of-flight mass spectrometers for GC, ground-breaking Select-eV ion-source technology and associated software.

**SHIMADZU**
Excellence in Science

Shimadzu Corporation is a Japanese company, manufacturing precision instruments, measuring instruments and medical equipment, based in Kyoto, Japan. The company was established by Genzo Shimadzu in 1875. Shimadzu UK covers the whole bandwidth of analytical instrumentation consisting of chromatography (HPLC, GC), mass spectrometry (LC-MS, GC-MS, MALDI-TOF MS), spectroscopy (UV/VIS, FTIR, AAS, ICP, EDX) and sum parameters (TOC). These product lines are complemented by weighing technologies, testing machines and the biotechnology business division.

Shimadzu's analytical instruments are used in the chemical, pharmaceutical and biotechnology laboratories, in the food and semiconductor industry, in research and development as well as in process and quality control. Through our Head Office in Milton Keynes and our Scottish Office in Livingston we provide dedicated local sales & service support including installation, warranty repair, on-going annual preventative maintenance, emergency intervention services, applications and software support, readily available OEM parts and consumables, equipment qualification services and user-training courses tailored to meet your specific needs.

Thermo Fisher Scientific is the world leader in serving science, enabling you to make the world healthier, cleaner and safer. With annual sales of more than $17 billion and 50,000 staff, we serve our customers worldwide within a broad range of markets and sectors.

With four premier brands we help solve analytical challenges from routine testing to complex research and discovery. Thermo Scientific is the leading brand for technology innovation and offers customers a complete range of high-end analytical instruments as well as laboratory equipment, software, services, consumables and reagents to enable integrated laboratory workflow solutions across all fields.

Welcome to JSB!

Over the years we have gained knowledge and experience to provide laboratory with chromatography solutions in the following areas: environmental, petrochemical, chemical, food & flavour, life sciences and industrial. Since 2002 JSB has been active in the European market and for the past ten years JSB has positioned itself as a full knowledge partner in chromatography and mass spectrometry solutions. Your problem is our inspiration, which often results in innovative solutions. We think very highly of both knowledge and innovation. if no commercial solution is available JSB will produce a solution for you in our workshop. JSB bases its systems on the GC (MS) and LC (MS) modules of Agilent Technologies.

Together with the large variety of sample pre-processing and detection equipment from several partners such as AFP, CTC, CDS, EST, IonSense, Owlstone, SAI and Zoex, JSB offers a full range of analysers and accessories. By means of our own research and development department solutions can be combined, modified or developed.

The Royal Society of Chemistry is the world's leading chemistry community, advancing excellence in the chemical sciences. With 48,000 members and a knowledge business that spans the globe, we are the UK's professional body for chemical scientists; a not-for-profit organisation with 170 years of history and an international vision for the future. We promote, support and celebrate chemistry. We work to shape the future of the chemical sciences – for the benefit of science and humanity.

# Contents

| | |
|---|---|
| Interaction between the regulator and practitioners in the field of environmental investigations<br>*Anne Brosnan* | 1 |
| An introduction to a multiparameter approach to improve the reliability of environmental crime evidence<br>*Cristina Barbieri, Jorge Eduardo Souza Sarkis, Marcos Scapin, Maria do Carmo Ruaro Peralba, Luiz Martinelli and Marcos Antônio Hortellan* | 7 |
| Environmental litigation issues in Taiwan and the forensics strategies of chlorinated hydrocarbon contaminated sites<br>*H.L. Hsu, P.H. Liu, H.C. Hung, and F.C. Chang* | 22 |
| Forensic investigations in a PCB case: potential pitfalls and tips<br>*Jean Christophe Balouet, Francis Gallion, Jacques Martelain, David Megson and Gwen O'Sullivan* | 31 |
| Use of symmetric tetrachloroethane to age date chlorinated solvent releases<br>*Robert D. Morrison* | 41 |
| A forensic methodology for investigating chlorinated solvent releases from a dry cleaner<br>*Robert D. Morrison* | 51 |
| $^{13}$C and $\delta^{37}$Cl on gas-phase TCE for source identification investigation - Innovative solvent-based sampling method<br>*Daniel Bouchard, Patrick W. McLoughlin, Daniel Hunkeler and Robert J. Pirkle* | 70 |
| Characterisation and modeling of dust in a semi-arid construction environment<br>*John Bruce, Hugh Datson, Jim Smith and Mike Fowler* | 82 |
| Assessment of polycyclic aromatic hydrocarbons in an urban soil dataset<br>*Rory Doherty, R. McIlwaine, L. McAnallen and S. Cox* | 92 |
| Statistics of lost historical coal tar PAH contamination in soils and sediments<br>*Michael J. Wade* | 104 |
| Micro-bubbles oxygen injection in groundwater contaminated by organic biodegradable compounds and metals<br>*A. D'Anna, R. Brutti, A. Gigliuto, R. Vaccari, G. Bissolotti, E. Pasinetti and M. Peroni* | 130 |

Heavy metals (lead, cadmium, chromium, nickel and zinc) contamination in selected electronic waste dumpsites in Lagos, Nigeria 138
*V.F. Doherty, M. K. Ladipo and I. O. Famodun*

Evidence for acid mine drainage in an urban stream in the West Midands 152
*Stephanie M. Turnbull and C.V.A Duke*

A Practical Evaluation of Petrogenic and Biogenic Methane Sources in the Context of Redevelopment and Explosive Hazard Management in Los Angeles, California 162
*Leo M. Rebele and Michael J. Crews*

Author Index 168

Subject Index 169

INTERACTION BETWEEN THE REGULATOR AND PRACTITIONERS IN THE FIELD OF ENVIRONMENTAL INVESTIGATIONS

Anne Brosnan

Deputy Director (Chief Prosecutor) Legal Services, The Environment Agency, England

1. INTRODUCTION

Prosecution of environmental offences almost invariably involves the presentation of complex evidence to a court, dealing with the causation of an offence, and its implications, in terms of harm or potential for harm to the environment, to flora and fauna, and to human health. As the principal environmental regulator for England, the Environment Agency issues environmental permits and then goes on to monitor, investigate and prosecute breaches. Additionally the Agency prosecutes environmental offences where there may be whole scale illegal activity outside of the regulatory regime. The Agency relies upon evidence to set its permits and subsequently to justify interventions in situations where operations may have an adverse impact on the environment or are already having such effects.
 The importance of sound, reliable evidence cannot be understated for a regulator. In cases brought to court by the Agency and in other challenges relating to the potential for harm associated with a proposed operation, the use of expert evidence and expert witnesses is crucial. Cases may require interpretation of the analysis of water and waste samples and what these mean in terms of environmental impact. Expert witnesses are frequently called upon to advise during investigations and to give evidence on complex issues to a court or tribunal.
 The Environment Agency utilises many analytical techniques, through the services of its own National Laboratory Service, but has additionally used chemical fingerprinting of petroleum products in cases such as the explosion at the Buncefield Oil Storage Terminal and other cases relating to losses from underground petrol and diesel storage facilities. Cases have been brought by the Agency in relation to the loss of pesticides, fertilisers, food products and Tributyltin (TBT). The impacts on human health but also on flora and fauna, fish and shellfish must be assessed and verified. Accordingly a good relationship between the regulator and the scientist is key and the discipline of environmental forensics is increasingly coming to the fore. Interaction between different professionals, who may not always speak a common language, is critical and we will consider how this can best be facilitated, with case related examples.

**Figure 1** *Buncefield explosion which caused pollution of groundwater*

The Environment Agency (the EA) is the largest of the UK's environmental regulators, established by statute in 1995.[1] It seeks to control polluting activities and operations and those with potential to cause harm to human health and the environment.

It is the principal aim of the Agency to protect or enhance the environment, taken as a whole, so as to make a contribution towards attaining the objective of achieving sustainable development taking into account any likely costs involved. [2]

The principle of cost-effective regulation is restated in the directly applicable 'Regulators' Code', the Statutory Code of Practice for Regulators.[3] Regulators must generally have regard to the Code when developing policies and operational procedures that guide their regulatory activities. This formally establishes the principle of risk based regulation. Section 3, headed 'Regulators should base their regulatory activities on risk' states:-

3.1 Regulators should take an evidence based approach to determining the priority risks in their area of responsibility, and should allocate resources where they would be most effective in addressing those priority risks.

The proposition that resources should be targeted where they will be most effective is not new but is explicitly stated here by Government. This principle is well expressed in the Hampton Report which concluded: [4]

"Risk assessment involves the identification and measurement of capacity to harm and, if such capacity exists, an evaluation of the likelihood of the occurrence of the harm. By basing their regulatory work on an assessment of the risks to regulatory outcomes, regulators are able to target their resources where they will be most effective and where risk is highest."

The EA is a permitting authority, an operating body and a regulator. As a modern regulator, the EA seeks to minimise unnecessary burdens by taking a purposive approach to permitting. After the permitting stage, compliance is assured in most cases by some degree of monitoring and enforcement. Much of the environmental protection legislation which it enforces comes from Europe, in particular Directive 2008/99 on the Protection of the environment through criminal law requires member states of the EU to have criminal penalties available for serious infringements of community law on protection of the environment. This criminalises behaviour which is damaging or potentially damaging to the environment. Because of the specialist nature of this work, the EA has an in house enforcement and prosecution capability. It employs environmental specialists who will consider the best interests of the environment throughout the duration of an operation which may potentially cause harm or risk to human health. The Agency employs its own scientists who help to set the parameters within which operators must function and agrees industry norms. It does not however retain all the experts it might need in complex and challenging enforcement situations and will in many cases bolster its own expertise with the use of external experts.

The Agency has enforcement priorities (including for example waste crime) which have, of necessity, become increasingly more sophisticated over time as it has had to manage competing demands across its many and varied functions. It has a small dedicated team of environmental investigators, an in-house intelligence team and accredited financial investigators. This is still a very minor part of overall regulatory activity but absolutely crucial to successful regulation and worthy of further consideration.

## 2. ENVIRONMENTAL INVESTIGATIONS

An environmental investigation may seek to establish the cause of a pollution incident, the extent to which authorised discharge levels have been exceeded or whether or not the risk of pollution or harm to human health is being or has been properly managed. Such an investigation must be conducted to the standards of an ordinary criminal investigation. Accordingly interviews are conducted subject to the relevant rules and safeguards. [5,6] In most cases an investigation will additionally however require on site investigations and the obtaining of sample evidence. This may be samples of materials entering a waterbody, materials deposited or released into the environment, samples taken to establish background levels of pollutants, control samples and those which are taken to determine impact upon the receiving environment. This is where the regulator turns to the discipline of environmental forensics for assistance.

## 3. FORENSIC INVESTIGATION

The purpose of a forensic investigation is to obtain sufficient reliable admissible evidence which can be used in formal proceedings or underpin any softer regulatory approach to suspected illegal activity or non compliance. It is vital that lawyers and scientists can work together and understand each other's terminology and disciplines. This does require lawyers to understand a little of the science and for scientists to understand something of the rules under which lawyers must operate.

## 4. SAMPLE EVIDENCE

Historically samples were taken pursuant to a very prescriptive process for tripartite sampling.[7] Such requirements have been repealed over time and there is no longer a requirement for the retention of a third portion of a sample to determine disputes between the regulator and third party. However samples are invariably taken in enforcement situations, usually subject to a formal process which involves service upon a potential defendant, arrangements to allow for continuity or chain of custody evidence and analysis under an accredited regime. Often it is necessary to examine background levels of pollutants and to establish the pre existing condition of a site. It is important that sample evidence produces data which are fully representative, in terms of both qualitative and quantitative analysis, of the materials sampled in the field. In some cases chemical fingerprinting techniques may be used for example where there has been a large scale spill of hydrocarbons and the origins of these need to be established as in the Buncefield case.

## 5. ACCREDITATION AND ANALYSIS

It is essential in the presentation of all scientific evidence that analysis is undertaken in accordance with the strict requirements of an accredited scheme such as UKAS accreditation.[8] This provides confidence in data achieved through analysis as the methods utilised during the analysis will be to established standards in relation to the handling transport and analysis of materials.

In a recent case, the sample of a pesticide spill running into a surface water ditch was broken in transit. The EA instructed an expert witness who was able to interpret all of the downstream results and extrapolate these into a report proving the entry of polluting materials at a point where a sprayer had overturned and discounting or explaining a rogue spike in the gas chromatography analysis, which was due to a mix of chemicals being used in the field spraying equipment.

**Figure 2** *Incident involving surface water pollution from chemical sprayer*

## 6. EXPERT WITNESS EVIDENCE

Environmental cases will often involve evidence being adduced with which courts may be unfamiliar. The effects of a pollutant in the water environment can be catastrophic and the court will need to understand something of the biology and complex ecosystems involved when handling these cases. This will therefore invariably involve reliance upon the evidence of an expert witness.

In England and Wales the rules governing the use of expert evidence are found in Rule 33 Criminal Procedure Rules: See below. The prosecutor will manage the use and instruction of expert witnesses in order to ensure that an expert witness understands the basis upon which he is instructed and his overriding responsibilities to the court.

The EA will use both in house experts and external expert witnesses, such as experienced academics or consultants depending on the requirements of the case. All experts called to give evidence must have the necessary skills and experience to be persuasive to the courts and must be demonstrably objective. The choice of expert witness is a key decision and is often critical to the success or failure of a case. The EA liaises with professional associations to ensure the highest level of critical appraisal is brought to bear upon its cases, both prior to commencement in court and then to assist the court in determining the relevant issues. The requirements governing the use of expert evidence are found at Rule 33 in the Criminal Procedure Rules.

### Rule 33.2 Criminal Procedure Rules —

(1) An expert must help the court to achieve the overriding objective [fairness] by giving objective, unbiased opinion on matters within his expertise.

(2) This duty overrides any obligation to the person from whom he receives instructions or by whom he is paid.

(3) This duty includes an obligation to inform all parties and the court if the expert's opinion changes from that contained in a report served as evidence or given in a statement.

### An expert's report must—

(a) give details of the expert's qualifications, relevant experience and accreditation;

(b) give details of any literature or other information which the expert has relied on in making the report;

(c) contain a statement setting out the substance of all facts given to the expert which are material to the opinions expressed in the report, or upon which those opinions are based;

(d) make clear which of the facts stated in the report are within the expert's own knowledge;

(e) say who carried out any examination, measurement, test or experiment which the expert has used for the report and—

   (i) give the qualifications, relevant experience and accreditation of that person,

(ii) say whether or not the examination, measurement, test or experiment was carried out under the expert's supervision, and
(iii) summarise the findings on which the expert relies;

(f) where there is a range of opinion on the matters dealt with in the report—
(i) summarise the range of opinion, and
(ii) give reasons for his own opinion;

(g) if the expert is not able to give his opinion without qualification, state the qualification;

(h) contain a summary of the conclusions reached;
(i) contain a statement that the expert understands his duty to the court, and has complied and will continue to comply with that duty; and

(j) contain the same declaration of truth as a witness statement.

## 7. VALUE OF FORENSIC ANALYSIS TO THE REGULATOR: CONCLUSION

It is imperative that good environmental regulation is underpinned by sound scientific evidence. In the enforcement arena where significant penalties and reputational damage can be occasioned by regulator interventions it is crucial that the regulator makes its decisions on the basis of good science. In turn this must be followed through to courts and tribunals with sufficient reliable relevant corroborative evidence to allow the courts to reach decisions with certainty and clarity. In a prosecution regime, or where civil sanctions are based upon the premise of an offence having been committed, as under the Regulatory Enforcement and Sanctions Act 2008, there is no room for error or ambiguity. The court must be satisfied beyond reasonable doubt of the validity of the prosecution case, including proof of the cause and nature of a pollution event or breach of permit. It will wish to hear further evidence as to impact and extent of any pollution arising including both harm and risk of harm. The credibility of the regulatory regime therefore rests upon the facts and the science behind regulator decisions. It is not possible to overstate the need for any regulator to call upon the best quality experts and scientific analysis to assist themselves and the courts and for the courts to understand, and to rely, with the utmost confidence, upon all of the experts and the evidence which is called before them.

## References

1. Environment Act 1995.
2. Environment Act 1995, ss.4.
3. The Legislative and Regulatory Reform Act 2006, ss.23 (effective April 2014).
4. P. Hampton, *Reducing Administrative Burdens: effective inspection and enforcement*, HM Treasury on behalf of the Controller of Her Majesty's Stationery Office, 2005.
5. Police and Criminal Evidence Act 1984.
6. Criminal Procedure and Investigations Act 1996.
7. Water Resources Act 1991, ss.209 (now repealed).
8. UKAS (United Kingdom Accreditation Service) www.ukas.com (accessed: February 2015).

# AN INTRODUCTION TO A MULTIPARAMETER APPROACH TO IMPROVE THE RELIABILITY OF ENVIRONMENTAL CRIME EVIDENCE

Cristina Barbieri[1,2], Jorge Eduardo Souza Sarkis[2], Marcos Scapin[2], Maria do Carmo Ruaro Peralba[3], Luiz Martinelli[4] and Marcos Antônio Hortellani[2]

[1] Centro de Química e Meio Ambiente, Instituto de Pesquisas Energéticas e Nucleares (IPEN),Cidade Universitária , São Paulo, SP , Brasil
[2] Instituto-Geral de Perícias, Secretaria de Segurança Pública do Rio Grande do Sul , Porto Alegre, RS , Brasil
[3] Departamento de Química, Universidade Federal do Rio Grande do Sul (UFRGS), Porto Alegre, RS, Brasil
[4] Laboratório de Ecologia Isotópica, CENA, Universidade de São Paulo (USP) , Piracicaba, SP, Brasil

## 1 INTRODUCTION

Obtaining sound evidence of environmental pollution crimes is usually challenging in environments impacted by multiple sources. This is particularly true when the situation involves running waters systems and the contaminants discharged by the suspect facility must be searched in stream-bed sediments, which are typically heterogeneous in terms of texture and composition. The heterogeneity reflects on the high variability of contaminant content, which implies that a large number of samples must be analyzed in order to reduce the uncertainty of the resulting data to an acceptable level. In environmental crimes investigations, where the State or "People" must provide the scientific proof, resources as laboratory analyses are few, and there are also time restraints, therefore, more effective and fit-for-purpose methodologies must be developed and tested in order to provide the justice operators reliable evidence in a feasible manner.

The present work introduces a case study of leachate discharge from a hazardous waste landfill in a heavily polluted watercourse where the use of a multiparameter approach using metals, stable isotope and organics analysis in stream-bed sediments was performed as an exploratory means provide evidence of this possible crime. A brief review of methods that explore multiple variables to add qualitative consistency for the results obtained is also presented.

Environmental crime is a somewhat new concept in international law. The notion of the environment as being an asset so important for society that it must be protected by the Penal Law is still under construction worldwide, since different approaches are used to its definition and enforcement. According to the European Commission, it *"covers acts that breach environmental legislation and cause significant harm or risk to the environment and human health"* (European Commission [1]). In Brazil, environmental crime definition is no different than the definition for other crimes. It is defined as any act that violates a criminal statute. This is possible because, in 1998, a specific statute that defines the acts subject to criminal litigation known as "Environmental Crime Act" was promulgated. The crimes are divided, in this statute, in five sections: the first are crimes against the fauna; the second, crimes against flora; the third pollution related crimes; the fourth crimes against public property (landscape, architectural, etc.); and the fifth are crimes against the environmental administration, those practiced by subjects within the framework of environmental permits system. This tactic is, in some ways, appropriate, because several criminal conducts need to be defined by other technical legislations and regulations or administrative provisions that can be more easily updated as required, once environmental law is a dynamic law, so new scientific findings, and experience from existing environmental law are used to improve legislation. Other advantage is that criminal conducts are well defined, so the prosecutor does not need to prove a criminal intent or if the conduct was or not a crime, just to prove that there was a crime and who was the responsible for it. Such a system of prosecution implies someone being charged as guilty for the criminal offence, therefore a suspect must be found and investigated in order to sustain a criminal litigation. The burden of proof imposed on the prosecution in the criminal law means that the State must provide the evidence to support the charges and convict the defendant beyond a reasonable doubt. Unlike the U.S. where the environmental crime investigation is usually performed by regulatory agencies, such as USEPA and state agencies, in Brazil it is performed by the Police and the evidence is provided by the Official Forensic Agencies.

The unlawful discharge of effluents in a watercourse, which is the theme of the present work, is one of the pollution related crimes defined in Brazilian Environmental Crime Act. The study case is about hazardous waste landfill whose manager was indicted for discharging untreated leachate in a small watercourse. The discharge occurred systematically, but was noticed after a massive fish die-off that occurred in drought conditions in Sinos River, where the affected stream discharged its waters.

In order to search for evidence of the unlawful leachate discharge from the landfill, which received mostly residues from tanneries and shoemaking factories, in the watercourse, stream-bed sediment samples were collected in the affected watercourse and analysis of metals were performed. These analyses did not provide a sound proof of the leachate discharge since there were others tanneries that discharged its effluents upstream the landfill. Further analyses were needed to distinguish the leachate from the other pollution sources, therefore isotope ratio analyses were performed in the samples.

The isotope analyses were selected because the consequence of the isotope fractionation processes that occur in the decomposition of organic waste is a distinctive isotopic signature of carbon from landfill leachate that can be used to identify leachate contamination in groundwater (Baedecker and Back [2]). North et al [3] have employed, besides the carbon isotopes in Dissolved Inorganic Carbon ($\delta^{13}$C-DIC), ammonia nitrogen isotope ratios ($\delta^{15}$N -NH$^4$) to detect the presence of leachate in watercourses associated with landfills concluding that isotopic measurements have the potential to be used as a tracer of leachate in surface waters.

The PAHs are one of the most important classes of anthropogenic persistent organic contaminants. They are relevant in pollution crimes investigation because many of these compounds are listed in group 2A (probable carcinogens) of the International Agency for Research on Cancer (IARC) and benzo(a)pyrene is included in the group 1 meaning that there is sufficient evidence of carcinogenicity in humans for the compound. PAHs are amongst the contaminants found in landfill leachates (Katkevičiūtė et al [4]; Zakaria [5]). To add up contaminant information to the previous data set, these compounds were determined in the stream sediments.

## 1.1 Study area

The watercourse that allegedly received the leachate discharges is Portão Stream. This watercourse runs through the urban area of two small cities, receiving discharges of domestic sewage and industrial effluents before the end of its course on the left bank of the Sinos River, 20 km downstream.

The study area encompasses approximately ten kilometers of Portão Stream, along which seven sampling sites were located (Fig. 1). One of these sites (Point 4) was located downstream some leachate outfalls of a landfill that was criminally charged for releasing untreated leachate directly into Portão Stream. In this landfill confined about 1,000,000m³ of industrial hazardous waste in the time of the fish die-off. The untreated effluent contained both particulate and dissolved phase contaminants.

The geological substrate of the studied area is predominantly sandstone belonging to the Pirambóia Formation of the São Bento group (Oliveira et al.[6], Machado and Freitas[7]). The major mineral constituent is quartz and the predominant grain size in sediments is sand (Robaina et al [8]).

## 2 MATERIALS AND METHODS

### 2.1 Sampling sites
Sampling points were recorded referenced to Datum SAD69, in UTM Zone 22J, using a Garmin GPS (Global Positioning System) receiver unit, model Etrex Summit HC, with estimated position error of less than five meters. Quantum GIS program, version 2.0.1, Dufour (Quantum GIS Development Team [9]) was used to the geoprocessing of spatial data and preparation of maps.

Seven sediment sampling sites were defined along Portão Stream course. The spatial distribution of the sampling sites is shown in Fig. 1 (Points 1 to 7). Point 1 is situated close to the headwaters of the stream. Nearby this site, there were no significant sources of pollution; therefore, this point was admitted as reference point for background levels. The points 2 to 7 are distributed downstream of the water course segment where there are major sources of pollution. In these samples, bulk carbon and nitrogen isotope ratios were determined as well as metals Cr, Cu, Pb and Zn were analyzed by Atomic Absorption Spectrometry.

Another sampling campaign was performed and material was collected in nine sampling sites. One of these sampling sites was Point 1 from the previous campaign, Points A to C were located upstream the landfill, Points D to F in the proximities of the landfill and points G and H in another stream located to the south of the landfill that was also expected to be affected by landfill discharges (Figure 2). This stream reaches Portão Stream downstream the main points of leachate discharge of the landfill. Point C is coincident with Point 2 from the previous campaign.

### 2.3 Metals
Sediment samples were analyzed for chromium (Cr), copper (Cu), lead (Pb) and zinc (Zn) concentrations in the Laboratory of Chemical and Isotopic Characterization (CQMA)/IPEN/USP. The dry sediment was acid digested using a Microwave Accelerated Reaction System, Model MARS 5®, according to recommendations of 3051A USEPA method. Metal concentrations were measured using Flame Atomic Absorption Spectrometer (FS-FAAS, Varian, model Spectr-AAS-220-FS).

All glassware was cleaned in 10% $HNO_3$ (w/v) prior to each experiment. Chemicals used for digestion and extraction experiments were analytical reagent grades. The validation of this method was performed by analyzing certificate reference sediments (SRM 2704 Buffalo River Sediment) in three replications.

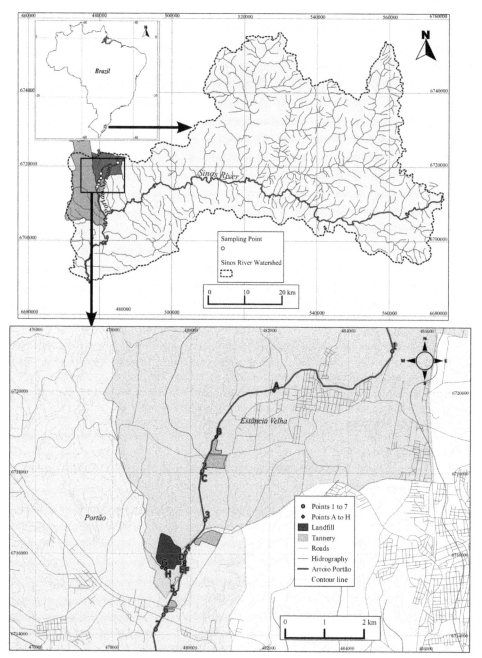

**Figure 1** *Study area location depicting the sampling points spatial distribution. Modified from Barbieri et al* [10]

**Figure 2.** Study area showing the sampling points located nearby the landfill

The samples collected in the second campaign were freeze dried and analyzed by Wavelength-Dispersive X-ray Fluorescence (WD-XRF) at X-ray Fluorescence Laboratory/ CQMA/IPEN/USP. The results obtained were expressed in mean values and standard deviations for a confidence interval of 95%. The detection limit of this method was 50 µg $g^{-1}$ for the metals Cr, Mn, Zn, Ni, Cu and 0,02% for the elements Si, Al, K, Fe, Mg, S, Na, Ti, P and Ca.

## 2.4 Carbon and Nitrogen Isotope Ratio

The isotopic compositions of total carbon ($\delta^{13}C$) and nitrogen ($\delta^{15}N$) in sediments were determined in samples by on-line automated combustion coupled to mass spectrometer Finnigan MAT Delta-S. The $^{13}C/^{12}C$ ratio is reported in relation to the VPDB standard carbon dioxide from calcium carbonate from the Pee Dee Belemnite formation, by convention, in $\delta^{13}C$ units per mil (‰). The precision of this analysis is 0.3‰. The $^{15}N/^{14}N$ ratio is reported in relation to the atmospheric air in $\delta^{15}N$ units per mil (‰) and the precision of this analysis is 0.5‰.

## 2.5 Polycyclic aromatic hydrocarbons (16-PAH)

About 500 g of sediments were collected from 0 to 30 cm of depth in a core like device, which was thoroughly rinsed with the stream water and after with distilled water prior to each sampling. The outer layer of the sediment was discarded and the inner part was stored in aluminum foil trays pre-heated to 300°C overnight and freeze stored until analysis. The samples where then freeze dried, a 20g fraction of the homogenized sample was added to a pre-extracted cartridge and extracted in with dichloromethane in an Automated Soxhlet System (SOXTEC) for 4 h using the configurations provided by the equipment manufacturer. The extract was concentrated in a rotary evaporator until approximately 1 mL, passed through an activated copper column, concentrated to 1 mL and fractioned by preparatory liquid chromatography in a silica column to obtain the aromatic fractions following the Jaffé et al. [11], method.

The fractions obtained were analyzed using a gas chromatograph Agilent (model 6890) with mass detector (model 5390), splitless injector and a capillary column (30 m X 25 mm X 25 μm) and stationary phase type 5% phenyldimetilpolisiloxane. Injector temperature: 290 °C; detector temperature: 300 °C; column starting temperature 40 °C, isotherm for 1 min, heating rate of 6 °C /min until a final temperature of 290 °C and isotherm of 20 min. The conditions applied were according to the Environmental Protection Agency (EPA), using the technique of selective ion monitoring and electron impact at 70 eV for ionization, based on the methodology EPA 8270C. The results obtained for the 16 priority PAH were qualitative and are shown in terms of presence or absence of the detected compounds.

## 2.6 Descriptive and statistical treatment

Principal Components Analysis (PCA) was employed as an exploratory tool to distinguish sampling sites based on the variations observed in the metals and carbon and nitrogen isotopic ratios analyses. The results obtained in the first sampling campaign (points 1 to 7) were arranged in correlation matrices in two steps in each sampling campaign data. First, the PCA was employed with metals concentrations and with isotope ratios separately, and after, data from metals concentrations and isotope ratios was assembled in a single matrix and the PCA was performed. The same method was used with the results metal concentrations obtained from WD-XRF analyses in samples collected in the second sampling campaign. The elements Al, Si, K, Fe, Mg, Na, P, Ca, Zr, Rb and Sr were not used in the PCA analysis because it was assumed that some of these elements were mostly associated with the sediments matrix or they were not detected in any of the sampling sites. The statistical analyses were performed using the PAST package, version 3.0 (Hammer [12]) and the graphical outputs obtained were edited in the vector drawing program Inkscape ©, version 0.48.

3 RESULTS AND DISCUSSION

The results obtained in the two sampling campaigns will be presented separately and a discussion of both approaches as well as a brief review of strategies revealed by other authors that could be applied to additional discussion of the criminal environmental forensic methodologies to be developed.

## 3.1 Metal Concentration (FS-AAS)

The results obtained in the analysis of Cr, Cu, Pb and Zn by FS-AAS in sediment samples are shown in Table 1. Point 4, in accordance to historical data from pollution monitoring of the stream, presented in Barbieri et al [10], showed the highest concentrations for all metals and very high concentrations of Cr. The metals analyzed did not exhibit increase trend in their concentrations along the watercourse. Relatively high Cr and Zn concentrations were also determined in Point 2. Metal concentrations were relatively higher in the sampling site located downstream the points of leachate discharge from the landfill (Point 4), comparing

**Table 1** *Metal concentration obtained by AAS analysis and isotope ratios of C and N in the sediment samples collected in Points 1 to 7.*

| Sampling Point | Cr µg g$^{-1}$ | Cu µg g$^{-1}$ | Pb µg g$^{-1}$ | Zn µg g$^{-1}$ | δ$^{15}$N (‰) | δ$^{13}$C (‰) |
|---|---|---|---|---|---|---|
| 1 | 30.40 | 12.79 | 5.56 | 19.42 | 7.06 | -22.15 |
| 2 | 437.61 | 10.10 | 1.01 | 43.10 | 3.75 | -23.69 |
| 3 | 85.85 | 14.68 | 7.50 | 32.74 | 4.13 | -24.32 |
| 4 | 953.89 | 17.87 | 21.68 | 50.87 | 6.05 | -23.43 |
| 5 | 61.35 | 10.79 | 7.26 | 26.48 | 4.00 | -24.85 |
| 6 | 22.46 | 9.29 | 10.23 | 17.31 | 3.88 | -26.33 |
| 7 | 12.15 | 10.37 | 8.29 | 35.05 | 3.32 | -27.92 |

to the other samplings sites, so that the PCA analysis allowed to distinguish this Point from the others (Fig. 3). Since there is also other possible sources of these metals upstream this point, the results *per se* could not point to a contamination specific from the suspect source.

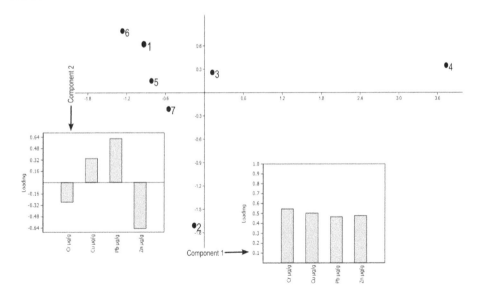

**Figure 3** *Scatter plot obtained from Principal Component Analysis of Cr, Cu., Pb and Zn, in the sampling points 1 to 7, showing the contribution of variables for each component. PC1 explains 73% of the variance in the data and PC2 explains 17%.*

## 3.2 Concentration and isotopic ratios of carbon and nitrogen

As presented in Barbieri et al [10], the results of the analyses of organic carbon isotope ratio ($\delta^{13}$C ‰) show a gradient of depletion in the heavier isotope from the source to the mouth of the watercourse by observing a variation of -22.2 ‰ to -27.9 ‰ from Point 1 to 7. This trend is not maintained in Point 4, which displays a greater enrichment in the heavier isotope in relation to previous and subsequent points in terms of spatial distribution. This enrichment in the heavier isotope noted in Point 4, could be explained by the inputs of a source enriched in $\delta^{13}$C in that region, such as landfill leachate which $\delta^{13}$C values can vary from -9.44‰ to 10.6‰, like demonstrated by Mostapa et al [13]. In the scatter plot obtained from the PCA analysis of these results (Fig. 4), Point 4 is also distinguished from the others. Point 1, used as reference site regarding industrial and domestic sewage contamination in the stream showed the values most enriched in $^{15}$N and the least enriched in $^{13}$C. This could explain why it loaded positively in PC1.

The $\delta^{15}$N values of sediment samples showed a general trend similar to that observed with carbon isotopes, that is to say, depletion in the heavier isotope from upstream to downstream. This trend, similar to carbon, was discontinued in point 4, which showed a relatively higher enrichment in the heavier isotope. Distinct and identifiable signatures of $\delta^{15}$N were determined for fertilizers and sewage, allowing nitrogen isotopic ratios to be successfully used in discrimination between these sources of contamination of aquifers, estuaries and oceanic environments (Andrews et al [14], Aravena et al [15], Rogers [16], Wassenaar [17]). Sewage derived organic matter presents $\delta^{15}$N values in the range of 3 ‰ (North et al [3]), which is compatible with those found in Points 2, 3, 5 6 and 7 (Table 1).

In the mineralization of organic matter that takes place in landfills organic nitrogen is converted into ammonia leading to a small change in the nitrogen isotope ratios (Kendall[18]). The volatilization of the ammonia causes an isotopic fractionation, leaving the remaining N highly enriched in $^{15}$N-NH$_4$ (North et al [3]), which could explain the values found for the Point 4. The high enrichment in the $^{15}$N-NH$_4$ of landfill leachate could cause the relative increase in the heavy isotope observed in Point 4, which lies in the area of influence of the landfill suspect of leachate discharge.

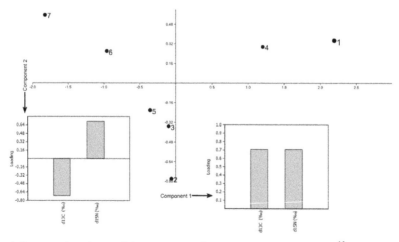

**Figure 4** *Scatter plot obtained from Principal Component Analysis of $\delta^{13}$C ‰ and $\delta^{15}$N‰ in the sampling points 1 to 7, showing the contribution of variables for each component. PC1 explains 89% of the variance in the data and PC2 explains 11%.*

## 3.3 Metal Concentration (WD-XRF)

The results obtained from WD-XRF analysis on sediments samples are shown in Table 2. In the displayed results, Points A and C show values above the detection limits for the metals Cr, Mn, Zn, Ni and Cu; Points B and F for Cr, Mn and Zn; Point E for Zn, Ni and Cu; Point G, that is located in the Boa Vista stream, for Cr, Ni and Cu. Point H, also situated in Boa Vista stream, did not show any metal concentration above the detection limit of the method. Point 1, which was used as a reference because it was situated upstream from industrial and domestic sewage discharges in the stream, presented the highest concentration of Mn. This could be due to the extensive cattle breeding around the sampling site, as livestock nutritional supplements usually contain Manganese (Nadaska[19]). Point D, although positioned in the segment of the stream likely to be affected by leachate discharges, only exhibited Mn concentrations above the detection limit of the method.

This analytical method was chosen because of the lower effort in sample preparation, comparing to AAS or ICP-MS, and to assess its usefulness in terms of fitness-for purpose as discussed in Ramsey [20], in this kind of situation. The low sensitivity of this method is not expected to be a problem in the greatest part of criminal environmental pollution investigations once most minimum regulatory values for metal contamination in soils, the exception is Ni in Brazilian regulation, are above the detection limit of the method. One limitation of the method used is that it did not include the metals Pb and Cd that have high capacity for accumulation in the environment, disturbing the biosphere equilibrium (Newman and Clements [21]), thus are elements of concern in pollution crime situations. The results of this analysis in the sediment samples were used to obtain the PCA scatter plot presented in Fig. 5. In this graph, Point C loaded positively on PC1 and PC2, since it was the one that presented higher concentrations for most elements while Points 1 and D loaded negatively in this axis. In both sites only Mn was detected. Points D, E and F, located in the area of influence of the landfill discharges showed different behaviors regarding metal contamination, as well as Point G and H, that were located in another watercourse that was also topographically related to the landfill. In the sample collected at Point D only manganese was determined above the detection limit and, in the sample collected at Point H, none of the heavier metals detected was identified. Points A and C, located upstream the landfill, but in industrial effluents polluted area, showed high concentrations of Cr, Mn, Zn, Ni, Cu. Among the points located in the area in the area of influence of the landfill discharges, Point F showed the higher concentrations of Cr and Zn.

The PCA analysis scatter plot displayed in Fig.5 does not discriminate the sampling points according to the inferred pollution sources based on XRF metal determination in sediments, once Point 1 is upstream industrial effluents discharges, points A, B and C are under industrial effluents discharge influence, and points D, E and F, under the landfill effluents discharge, but downstream A, B, and C. Points G and H were also located nearby each other and under the same macro influences in terms of metal contamination deposition.

## 3.4 Polycyclic aromatic hydrocarbons (16-PAH)

PAHs (Naphthalene, Phenanthrene, Fluoranthene, Pyrene, Chrysene, Benzo [k]fluoranthene, Benzo [a]pyrene, Benzo [g.h.i]perylene, Indeno [1,2,3-cd]pyrene) were detected in Points D, E, F and H as seen in Table 3. The spatial distribution of the sampling sites where these compounds were detected is topographically associated with the landfill. In sampling sites that showed high levels of metal contamination related with industrial effluents, located upstream the landfill, these compounds were not identified. Although

**Table 2** Metal concentration obtained by WD-XRF analysis in the sediment samples collected in Points 1 and A to H.

| | 1 | A | B | C | D | E | F | G | H |
|---|---|---|---|---|---|---|---|---|---|
| Si (%) | 32±1 | 39±1 | 39±1 | 37±1 | 37±1 | 41±1 | 41±1 | 43±1 | 42±1 |
| Al (%) | 5,5±0,1 | 4,6±0,1 | 5,0±0,1 | 4,8±1 | 6,3±0,1 | 3,9±0,1 | 3,0±1 | 2,6±0,1 | 2,8±0,1 |
| K (%) | 0,6±0,1 | 1,7±0,1 | 1,5±0,1 | 1,0±0,1 | 1,6±0,1 | 0,8±0,1 | 0,9±0,1 | 0,8±0,1 | 1,0±0,1 |
| Fe (%) | 4,4±0,1 | 1,6±0,1 | 1,5±0,1 | 2,1±0,1 | 1,5±0,1 | 0,9±0,1 | 1,0±0,1 | 0,6±0,1 | 0,5±0,1 |
| Mg (%) | 2,1±0,1 | 0,23±0,05 | 0,21±0,05 | 0,23±0,05 | 0,42±0,05 | 0,17±0,05 | 0,23±0,05 | 0,10±0,05 | 0,16±0,05 |
| S (%) | 0,28±0,05 | 0,30±0,05 | 0,32±0,05 | 0,64±0,05 | 1,3±0,1 | 0,26±0,05 | 0,29±0,05 | 0,14±0,05 | 0,09±0,05 |
| Na (%) | <0,02 | 0,17±0,05 | <0,02 | 0,14±0,05 | <0,02 | <0,02 | <0,02 | <0,02 | <0,02 |
| Ti (%) | 0,40±0,05 | 0,23±0,05 | 0,32±0,05 | 0,34±0,05 | 0,23±0,05 | 0,19±0,05 | 0,13±0,05 | 0,09±0,05 | 0,10±0,05 |
| P (%) | 0,07±0,01 | 0,08±0,01 | 0,10±0,05 | 0,35±0,05 | 0,08±0,01 | 0,14±0,05 | 0,17±0,05 | 0,09±0,01 | 0,06±0,01 |
| Ca (%) | 1,5±0,1 | 0,36±0,05 | 0,41±0,05 | 0,47±0,05 | 0,37±0,05 | 0,25±0,05 | 0,55±0,05 | 0,19±0,05 | 0,06±0,01 |
| Zr (µg g-1) | <50 | <50 | <50 | <50 | 0,007 | <50 | <50 | <50 | <50 |
| Rb (µg g-1) | <50 | <50 | <50 | <50 | <50 | <50 | <50 | <50 | <50 |
| Sr (µg g-1) | <50 | <50 | <50 | <50 | <50 | <50 | <50 | <50 | <50 |
| Cr (µg g-1) | <50 | 247±30 | 242±30 | 548±50 | <50 | <50 | 455±50 | 270±30 | <50 |
| Mn (µg g-1) | 1355±100 | 153±20 | 160±20 | 221±20 | 206±20 | <50 | 154±20 | <50 | <50 |
| Zn (µg g-1) | <50 | 530±50 | 498±50 | 131±20 | <50 | 530±50 | 562±50 | <50 | <50 |
| Ni (µg g-1) | <50 | 105±10 | <50 | 196±20 | <50 | 54±10 | <50 | 75±10 | <50 |
| Cu (µg g-1) | <50 | 106±10 | <50 | 199±20 | <50 | 55±10 | <50 | 76±10 | <50 |

these compounds were identified in points within the area impacted by the supposed leachate discharge, in Point G, that was located near point H, none of the compounds were detected. In Point D sample, which was collected close to Points E and F, only Perylene, a non-anthropogenic PAH (Ruffino et al [22]) was identified.

### 3.5 Multiparameter approach

The presented results of two sampling campaigns were used to illustrate a criminal environmental forensics procedure where budget and time restrains are frequently, if not always, important issues to be dealt with. This implies that the forensic examiner must make choices and seek alternatives in order to obtain relevant information within this limited framework.

In the case studied, the monitoring of stream pollution was made using metal concentration data which could be valuable for fingerprinting of sources (Yay et al [23], Gu et al [24]). However, usually a high number of samples are needed for these analyses to be statistically significant. The assemblage of data from additional parameters such as isotope ratio, reported as potential trackers of landfill leachate (North et al [3], North et al [25]; Mostapa et al [13]) and PAH analyses (Gade et al [25], Sisinno et al [27]), which are also used for

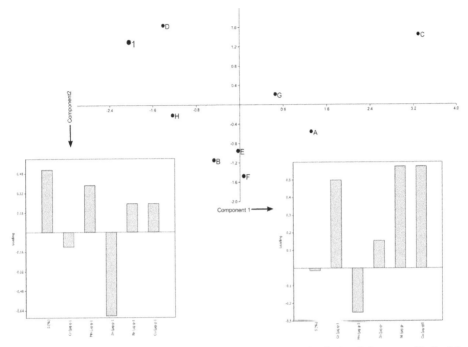

**Figure 5** *Scatter plot obtained from Principal Component Analysis of elements (S, Cr, Mn, Zn, Ni, Cu) concentrations analyzed by WD-XRF in the sampling points 1 and A to H, showing the contribution of variables for each component. PC1 explains 44% of the variance in the data and PC2 explains 23%.*

**Table 3** *PAH presence[1] in the sediment samples collected in Points 1 and A to H.*

|   | Nap | Phe | Flt | Pyr | Chry | Bkf | Bap | BghiP | Ind |
|---|---|---|---|---|---|---|---|---|---|
| 1 | - | - | - | - | - | - | - | - | - |
| A | - | - | - | - | - | - | - | - | - |
| B | - | - | - | - | - | - | - | - | - |
| C | - | - | - | - | - | - | - | - | - |
| D | - | - | - | + | - | - | - | - | - |
| E | - | + | + | + | - | + | - | - | + |
| F | + | + | + | + | + | + | + | + | + |
| G | - | - | - | - | - | - | - | - | - |
| H | + | + | + | + | + | + | + | + | - |

[1] + detected and – not detected
Nap: Naphthalene; Phe: Phenanthrene; Flt: Fluoranthene; Pyr: Pyrene; Chry Chrysene; Bkf: Benzo [k]fluoranthene; Bap: Benzo [a]pyrene; BghiP: Benzo [g.h.i]perylene; Ind: Indeno [1,2,3-cd]pyrene;

fingerprinting and source identification, should be evaluated as a potential tool for improving the reliability of pollution evidence and source identification.

The forensic use of a multiparameter approach to identify sources of contaminants could lead to a more accurate identification a suspect source. In this work, a better discrimination of the sampling sites according to the suspect sources is shown in the PCA scatter plot (Fig. 6) where metal concentrations and isotope ratio values for sediments were drawn together for the PCA analysis, using a correlation matrix to standardize the values that are in different scales. In this graph, Point 4, related to the landfill discharges, loaded positively in PC1 and was distinctly separated from other Points. Likewise, Point 1, considered as reference site because it was positioned upstream industrial and domestic sewage releases, loaded positively in PC2, distant from the others. Points 2, 3 and 5, that were under the influence of similar multiple sources, grouped together along the PC1 axis and Points 6 and 7, that were farther from the sources, also grouped. Even though the graph seems to indicate a solid distinction between points regarding the sources, this study was merely exploratory and additional analyses, with larger datasets, are needed to validate this approach as a forensic methodology.

The additional analyses performed in the sediments of the same stream and in another stream that was also topographically related to the landfill in the second sampling campaign, which included PAH and metals determinations, displayed fuzzier data.

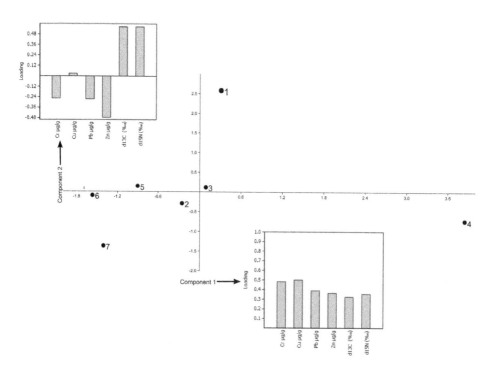

**Figure 6** *Scatter plot obtained from Principal Component Analysis of Cr, Cu., Pb, Zn, $\delta^{13}C$ ‰ and $\delta^{15}N$‰ in the sampling points 1 to 7, showing the contribution of variables for each component. PC1 explains 56% of the variance in the data and PC2 explains 25%.*

The PCA scatter plot of the results obtained in the metals analyses by WD-XRF in the samples collected does not discriminate the sampling sites based on the inferred pollution sources. This is probably due to the high detection limit of the method that excludes from the analyses the variations in concentrations below 50 µg g$^{-1}$, that are loaded in the dataset as "zeros", making this determinative method not suitable for obtaining metals levels to be used in a PCA analysis in this situation.

The qualitative PAH analysis added more information to the investigation because these compounds were detected only in regions supposedly affected by the landfill discharges. Comparing the results of the PAH and metals analyses, in the sampling points it can be seen that Point F is the only one that shows high levels of metals and all range of PAHs that were detected in these analyses. Point E also showed some metals and some PAHs. Points G and H, that were located nearby each other and in a region apparently affected by the same sources, instead, presented uneven results with Point G showing high levels of some metals and no PAH detected and Point H showing some PAHs and no metals.

It is noticeable that there are differences in metals concentrations and PAHs detected even in samples collected in nearby points, such as G and H and D, E and F, for. The heterogeneity of the sediment compartment and small-scale patchiness of this media (Chapman et al [28], US Army Corps of Engineers [29], Hadley et al [30]) are the probable causes of these apparently inconsistent results.

In criminal environmental situations, it is essential to remember that the absence of evidence is not *necessarily* evidence of absence. The geochemical spatial heterogeneity of stream sediments can easily lead to a type 2 error, which, in this case, would be to reject the pollution hypothesis, with a single or few samples analyses. This could be solved with an extensive sampling plan reaching a sample size (*n*) capable to provide evidence that there is indeed contamination on the suspect site and with fingerprinting methods that allow the identification of the source (the author of the crime) with a reasonable accuracy. The burden of proof is on the side of the State (people) in criminal litigation, therefore the examiner must seek means to provide solid evidence, so that the law officers can conduct the case effectively and the actual offender be properly condemned.

The development of evaluation methodologies to be applied in environmental data that are able to improve the significance of the findings would aid the environmental crimes investigations and verdicts.

One example of these methodologies is the Triad approach described by Chapman et al [27] for evaluating and assessing pollution-induced degradation in sediments which is based on three components sediment chemistry, sediment bioassay, and in situ parameters such as benthic community structure. The information provided by each component is unique and complementary and the components are combined to provide comprehensive information that cannot be obtained by a single component analysis. Delconte et al [31] used a multi-tracer multi-isotope approach to identify and quantify inputs of $NO_3$ in groundwater. Recently, Yang et al [32] and Xue [33] used a Bayesian model that ran under the open source statistical software R, called stable isotope analysis in R (SIAR) to source apportionment and estimation of proportional contributions of sources of nitrate, respectively. These methods could be useful if explored because they could be applied to Bayesian interpretation methods now commonly applied to evidence in courts (Ehleringer[34]).

## 4 CONCLUSIONS

In the presented study case, the assemblage of the metals analyses by EAA and stable isotope ratio data in to perform a Principal Components Analysis, provided a better discrimination of sediment sampling points according to pollution sources comparing to the use of the same method in the separate datasets (metals and isotopes) and with the results of metal analyses by XRF. The determination of metals by XRF and the qualitative determination produced fuzzy results requiring additional investigations.

These results were not fit for PCA analysis in this case because of the non-continuous character of the binary data generated by the qualitative PAH analyses and the high number of undetected elements that resulted from the relative high level of detection of the XRF analyses performed.

The apparently inconsistent results of the XRF and PAH analysis was probably due to the high heterogeneity of the sediment and this brings a concern for environmental crimes reports based on analyses in this compartment.

The use of a multiparameter approach, incorporating quantitative PAHs data to the metals and isotope analyses results, should be further explored in situations as the case presented to assess its applicability to improve the reliability of the criminal evidence. This approach could be useful when there are few samples and the environmental forensics experts need to extract the higher amount of information of them. Additionally, other evaluation methodologies as well as methods that can reduce or inform the uncertainties generated by the sediments heterogeneity ought to be developed in order to bring more trustworthy pollution crime evidence to courts. This could aid the environmental pollution crimes to attain the importance they deserve in the Criminal Law system and not be regarded as lesser crimes.

**References**

1. European Comission. Directive 2008/99/ECOF The European Parliament and of the Council on the protection of the environment through criminal law, 2008, http://eur-lex.europa.eu/legal-content/EN/TXT/?uri=CELEX:32008L0099 (accessed in June, 2014).
2. M. J. Baedecker, and W. Back, *Ground Water*, vol. 17, n° 5, pp.429 -437.
3. J. C. North, R. D. Frew, and B. M. Peake, *Environ. Int.*, 2004, **30**, ch.5, pp. 631–637.
4. J. Katkevičiūtė, P. Bergqvist, and V. Kaunelienė, *Environmental Research Engineering and Management*, 2006, **vol.3**, n° 3, pp. 29–35.
5. M. P. Zakaria, K. H. Geik, W. Y. Lee, and R. Hayet, R., *Coastal Marine Science*, 2005, **29**, pp.116-123.
6. M. T. G. Oliveira, S. B. A. Rolim, P. C. Mello-Farias, A. Meneguzzi, and C. Lutckmeier, C., *Water Air Soil Pollut.*, 2008, **192**, pp.183–198 DOI 10.1007/s11270-008-9645-8.
7. J. L. F. Machado, and M. A. Freitas, 2005. Projeto Mapa hidrogeológico do Rio Grande do Sul: relatório final, http://www.sema.rs.gov.br/sema/html/hidrogeologico/Relatorio Final.pdf, (acessed July, 2007).
8. L.E.S. Robaina, M.L.L. Formoso, C.A.F. Pires, Re*vista do Instituto Geológico*, 2002, **23**, n. 2, p. 35-47.

9. Quantum GIS Development Team, version 1.7.0 –Wroclaw, Quantum GIS Geographic Information System, Open Source Geospatial Foundation Project, 2010.
10. C.B. Barbieri, J.E.S.Sarkis, L. A. Martinelli, I.C.A.C. Bordon, H. Mitteregger Jr, M. A. Hortellani, *Environ. Forensics*, 2014, **15**, ch.2, pp.134-146.
11. R. Jafeé, G.A. Wollf, A. C. Cabrera, and H. Carvajal-Chitty, *Geochim. Cosmochim. Ac*, 1995, **59**, pp. 4507-4522
12. Ø. Hammer, D.A.T. Harper, and P. D. Ryan, 2001, *Palaeontol. Electron.*, **4**, ch. 1 pp. 1- 9.
13. R. Mostapa, T. Abdul, I. Abustan, and N., *International Journal of Environmental Sciences*, 2011, **1**, no. 5, pp. 948–958.
14. J. E. Andrews, A. M. Greenaway and P. F. Dennis, *Estuar. Coastal Shelf.* **46**, pp.743–756.
15. R. Aravena, M. Evans and J. Cherry, Ground Water, 1993, vol. 31, ch.2, pp.180–186.
16. K. M. Rogers, *New Zeal J Mar Fresh*, 1999, **33**, pp.181–188.
17. Wassenaar L.I., *Appl Geochem*, 1995, **10**, pp.391–405.
18. C. Kendall, in Isotope tracers in catchment hydrology, ed. C. Kendall, J. J. McDonnell, Elsevier, Amsterdam, 1998, pp. 519– 576.
19. G. Nadaska, J. Lesny, and I. Michalik, Environmental Aspect of Manganese Chemistry, 2012, http://heja.szif.hu/ENV/ENV-100702-A/env100702a.pdf, (accessed in July, 2013).
20. M. H. Ramsey, and K. A. Boon, *Appl. Geochem.*, 2012, **27**, ch.5. pp. 969-976. ISSN 0883-2927
21. M.C. Newman, H. Clements, Ecotoxicology - A Comprehensive Treatment, CRC Press, Boca Raton, 2008, 878 pp.
22. B. Ruffino, M. C. Zanetti, and G. Genon, *Soil Sediment Contam.*, 2009, **18**, ch. 3, pp.328-344. DOI:10.1080/15320380902799342.
23. O. Yay, O. Alagha, and G. Tuncel. *J. Environ. Manage.*, 2008, **86**, no. 4, pp. 581–594, DOI:10.1016/j.jenvman12.032.
24. Y. G. Gu, Z. H. Wang, S. H. Lu, S. J. Jiang, D. H. Mu, and Y. H. Shu, Environ. Pollut. 2012, vol. 163, pp.248–255, DOI:10.1016/j.envpol.2011.12.041.
25. J. C. North, R. D. Frew, and R. Van Hale, *J. Geochem. Explor.*, 2006, **88**, pp. 49–53.
26. B. Gade, , M. Layh, H. Westermann, and N. Amsoneit. *Waste Manage. Res.* 1996, **14**, no. 6, pp. 553–569, DOI:10.1177/0734242X9601400604.
27. C. L. Sisinno, A. D. Pereira Netto, E. C. Rego, and G. S. Lima, Cad. Saúde Pública, 2003, vol. **19**, ch. 2, pp. 671-676.
28. P. M. Chapman, *Sci. Total Environ.*, 1990, **97/98**, pp. 815–825.
29. US Army Corps of Engineers. Implementation of Incremental Sampling (IS) of Soil for the Military Munitions Response Program, United States: Department of the Army, 2009.
30. P. W. Hadley, E. Crapps, and A. D. Hewitt. *Environ. Forensics* **12**, no. 4, pp. 312–318. DOI:10.1080/15275922.2011.622344.
31. C. A. Delconte, E. Sacchi, E. Racchetti, M. Bartoli, J. Mas-Pla, and V. Re. Sci. Total Environ., 2014, vol.466/67, pp.924-938, DOI:10.1016/j.scitotenv.2013.07.092.
32. L. Yang, J. Han, J. Xue, L. Zeng, J. Shi, L. Wu, and Y. Jiang. *Sci. Tot. Environ.* 2013, **463–64**, pp.340–347, DOI:10.1016/j.scitotenv.2013.06.021.
33. D. Xue, B. De Baets, O. Van Cleemput, C. Hennessy, M. Berglund, and P. Boeckx. *Environ. Poll.*, 2012, **161**,pp. 43–49, DOI:10.1016/j.envpol.2011.09.033.
34. R. Ehleringer, Fourth FIRMS Network Conference, Washington, 2010.

ENVIRONMENTAL LITIGATION ISSUES IN TAIWAN AND THE FORENSICS STRATEGIES OF CHLORINATED HYDROCARBON CONTAMINATED SITES

Hsin-Lan. Hsu[1], P.H. Liu[1], H.C. Hung[2], F.C. Chang[2]

[1] Industrial Technology and Research Institute, 321, Kuang Fu Rd. Sec.2, East District, Hsinchu 30011, Taiwan
[2] Environmental Protection Administration, Executive Yuan, 83, Zhonghua Rd. Sec. 1, Zhongzheng District, Taipei 10042, Taiwan

1 INTRODUCTION

The industrial activity in Taiwan can be dated back to the nineteen century and soared during the World War II when a lot of metal working and chemical manufacturing were developed to support the military. Since then, industrialization has been the national reconstruction emphasis. Export processing zones and industrial parks were built one after another with a variety of factories gathered intensively. With a powerful solvency, chlorinated organic solvents such as tetrachloroethylene or trichloroethylene (TCE) were used in many of these industries, but registration for handling or manufacturing these toxic chemicals was not mandatory until 1998. The widespread usage but lack of waste management has led to a fact that groundwater contaminated by chlorinated organic hydrocarbons is ubiquitous across the country.

In 2000, the Soil and Groundwater Pollution Remediation Act was promulgated. Based on the Act, a site has to be enlisted first before a proper pollution control measure can be enforced and liability be imposed. To enlist a site, evidence should be provided to demonstrate that the pollution source is clarified or is located. In the past decade, a local maximum of the contaminant concentration was used by the government as a proof when listing a site. However, this evidence is facing challenges in court. After the government lost the first case, the Anding site lawsuit in 2012, the site was delisted and other sites are now following the same argument to get delisted.

The argument that the government was not able to defend was that the local maximum of the concentration in the groundwater could be a result of contaminant accumulation from off-site sources. As the geology of alluvium in the west coast of Taiwan, where the majority of industries reside, is often heterogeneous with embedded clayey texture, contamination accumulation could indeed occur. In addition, most contamination occurred where the factories are clustered, conclusions based on loosely distributed monitoring wells sometimes may be fragile if the wells were not located on the preferred transport pathway of the pollutants.

In this study, we revisited this Anding site and adopted compound-specific isotope analysis (CSIA) to show the source relevance. CSIA has been shown to provide unique information that can distinguish different sources.[1-3] Effort was also made to evaluate the presence of dense non-aqueous phase liquids (DNAPLs). With real-time probing technology, high resolution pollution distribution can be visualized and used to narrow down the area of source investigation, allowing for a more effective targeted sampling procedure to be undertaken.

## 2 SITE BACKGROUND

The Anding site is located in southern Taiwan. It was used for plastic fabrication between 1980 and 1994, prepared for satellite production in 1996, and then abandoned since 1997 due to a fire. The factories nearby include electronic manufacturing, metal-working industries, polymer manufacturing, casters and wheels manufacturing, timber processing and food industries as shown in Figure 1. None of them, including the former factories at Anding site, had a formal record of TCE handling.

In 2009, the Environmental Protection Administration (EPA) selectively investigated the factories once registered for electronics production by installing simplified ground water monitoring wells. TCE concentrations up to 3.37 mg/L were detected in the groundwater at this site. Subsequently, ten further monitoring wells were installed at the site as well as in the vicinity and sampled between 2009 and 2010, giving a concentration profile showing a local maximum at the north-east corner of the site as shown in Figure 1. The screen of these wells opens between 4 m and 10 m below ground surface (bgs), coinciding with the screen of the well where the TCE concentration of 3.37 mg/L was detected. Cis-1,2-dichloroethene was also detected in some of the groundwater samples, suggesting TCE degradation.[4] Seventeen soil cores were sampled randomly over the site at

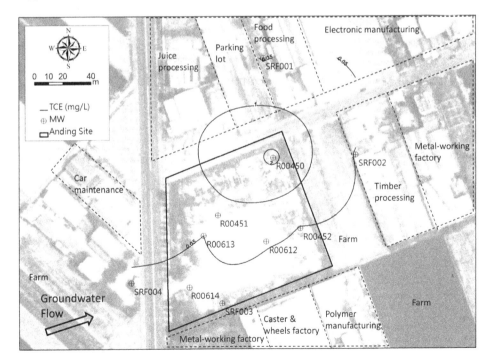

**Figure 1** *Anding site map showing the locations of monitoring wells, the contour of TCE concentration in groundwater, and the nearby land use (dash lines)*

depths ranging from 1.5 m to 11 m bgs. In the three samples taken at the depth between 7 m to 8 m bgs, higher TCE concentrations were detected ranging from 1.69 mg/kg to 6.32 mg/kg with an average of 4.40 mg/kg, while in the others TCE was either not detected or less than 0.4 mg/kg.

The geological formation of the site consists of a 10 m-thick layer of sandy loam overlying fine sands. No confining layer was observed within 15 m bgs. The organic carbon content of the sandy loam measured 0.38%. Depth to groundwater varies seasonally between 1.2 m and 3.5 m, with groundwater flow to east or east by north at an estimated rate of 0.23 m/yr in sandy loam. However, exceptions might occur during the typhoon season in the summer as this area is subject to flooding.

Since the investigation results in 2009-2010 showed that TCE was either not detected or detected at a low concentrations in the up-gradient monitoring wells and a local concentration maximum as indicated in Figure 1 by the closed contour was located at this site, the site was enlisted in 2011. Following the enlisting, an administrative lawsuit was filed by the landowner over insufficient evidence of pollution source at the site. The court decision was in favor of the plaintiff in that the possibility of mass accumulation of TCE migrating from up-gradient off-site sources was not excluded. Thus the site was delisted in 2013.

## 3 FORENSICS STRATEGY

To achieve a convincing determination of the pollution source, the forensics strategy comprised of two objectives. One is to examine the possibility of contaminant accumulation at this site and identify the number of sources. The other is to collect direct evidence of the source, such as the presence of DNAPLs and its spatial distribution that can be tracked back to the surface location where a spill could possibly have occurred.

To distinguish contaminant accumulation from a source, CSIA was employed. Owing to the kinetic isotope effect, TCE becomes enriched in the heavy isotopes as it degrades.[5] $\delta^{13}C$ and $\delta^{37}Cl$ values, thus increase as TCE migrates downstream. When the trend of isotope signature shows an abrupt change, particularly an inverse one, the presence of a source is signaled.[2,3] Such an argument has been used as a line of evidence to exclude the existence of a source at a site because the isotope signature did not show an abrupt change in the trend where a local concentration maximum was observed.[1]

From the same source, the isotope characteristics of the degraded TCE can be further quantitatively described by the Rayleigh equation (1):[6]

$$R = R_0 f^{\varepsilon/1000} \qquad (1)$$

where $R$ is the heavy-to-light isotope abundance ratio of remaining TCE, $R_o$ is the isotope abundance ratio of the TCE source, $f$ is the fraction of remaining TCE and is proportional to the TCE concentration, and $\varepsilon$ is the reaction-specific isotope enrichment factor. After transforming $R$ to the isotope signature notation $\delta$ (in per mil), the Rayleigh equation can be simplified, assuming the isotope signature is much smaller than unity, as below:

$$\delta = \delta_0 + \varepsilon \times \ln f \qquad (2)$$

where $\delta$ is the isotope signature of the remaining contaminants after degradation and $\delta_o$ is the isotope signature of the source. Deviation from the relationship may suggest the presence of other sources or other transformation pathways due to different cultures or geochemical environment.

After obtaining the number of the sources, investigating the presence of DNAPLs is often critical as it not only confirms that there had been a spill but also can possibly suggest the migration pathway of the spill from surface to the deep underground. There are several ways to show the presence of DNAPL, such as a groundwater sample showing an interface of water and organic liquid or the contaminated soil colored by a hydrophobic dye.[7] In order to collect the direct evidence in a cost-effective way, stable isotope signature was used to narrow down the target area of investigation. As the source tends to be enriched in the light isotopes compared to its downstream, the well whose $\delta^{13}C$ and $\delta^{37}Cl$ are smaller accompanied by a high contaminant concentration should be close or in the source area and thus was chosen as the starting point for the follow-up real-time probing. Locations of the probing were determined dynamically on site based on hydrology and the previous probing results. Compared with the slow, expensive but accurate laboratory analytical method, the use of probing tools such as membrane interface probe (MIP) is qualitative and not analyte-specific. Yet, it is a cost-effective way to identify where to concentrate the collection of soil core samples.

## 4 CHARACTERICATION AND ANALYTICAL METHODS

Characterization of TCE concentration and isotope signature was conducted at the monitoring wells where TCE was previously detected in the groundwater. The groundwater samples were packed on ice and shipped to the SGS laboratory (Taipei, TW) for volatile organic compound (VOC) analysis. The analysis was conducted in accordance with U.S. EPA SW-846 Method 8260B.

Groundwater samples for isotope analysis were transferred in part to another vial with headspace and equilibrated for at least two hours in the ambient. For carbon isotope analysis, compounds in the headspace were extracted for 40 minutes with a solid-phase-microextraction (SPME) fiber coated with carboxen and polydimethylsiloxane and then analyzed using gas chromatograph/combustion/isotope ratio mass spectrometer (GC/C/IRMS). The analytical system is composed of an Agilent 6890 GC, equipped with a DB-624 column for separation, a combustion furnace packed with Cu/Ni/Pt catalyst, and a ThermoFinnigan DeltaPlus XP IRMS. For chlorine isotope analysis, the compounds extracted from headspace were analyzed using gas chromatography with mass spectrometry (GC/MS) (Agilent 6890/Agilent 5975).[8] The individual abundance of the selected fragment ion, i.e., m/z = 60, 62, 95, 97, 130, and 132, was converted to derive the chlorine isotope ratio,[8] followed by calibration to the international standard mean ocean chloride scale (SMOC).[9] Triplicate analysis was carried out for each sample. Deviation of the δ values less than 0.5‰ for carbon isotope or 1‰ for chlorine isotope is within analytical tolerance in our method.

The real-time probing was performed using MIP in conjunction with a direct push platform. The sensors connecting to MIP were an electron capture detector (ECD) and a dry electrolytic conductivity detector (DELCD). Both respond to chlorinated organic compounds, but the latter is less sensitive and thus more indicative near the source when the ECD signal is approaching saturation. Results of probing along the same transect were extrapolated to generate a pollution map using Surfer 8.0 with ordinary kriging method.

To analyze the VOC concentrations in soil, the collected soil samples were extracted with methanol and then analyzed using GC/MS. To examine the presence of DNAPL, the tested sample was prepared by mixing approximately 20 g of the contaminated soil sample,

20 mL of water and 2 g of the hydrophobic dye, Sudan IV. This dye can stain NAPL to provide a visual contrast against the soil.

## 5 RESULTS

### 5.1 Differentiation of Sources and Mass Accumulation

Characterization results of composition and isotope signature of the groundwater were shown in Table 1. No TCE was detected at the outermost wells SRF001~SRF004. Among those which TCE was observed, both carbon and chlorine isotope values were higher in the up-gradient wells such as R00612 and R00613 and lower in the down-gradient wells such as R00450. If the contaminants were flowing from up-gradient off-site source and accumulated at the site, the variation trend of the isotope values would be the opposite. The depleted isotope signature at R00450 is also not likely due to the isotope fractionation effect of mass diffusion, which is often accompanied with relatively low concentration.[10] Thus the isotope data suggests that there is an onsite source of contamination which is contributing to the detected DNAPL.

The isotopic signature of samples obtained from different locations was further evaluated to see if there are signs of multiple sources. Both carbon and chlorine isotope values were well correlated with the TCE concentration by the Rayleigh equation within the analytical tolerance as shown in Figure 2. As a result, a second source of TCE was not suggested and the observed pollution is attributed to the same onsite source.

### 5.2 Evidence of the Presence of DNAPLs

As the isotope signature at R00450 is the most depleted among the five monitoring wells, the collection of direct evidence was then focused in its up-gradient area but not beyond R00451 and R00452. As of this point, information about contamination at the site was all obtained from the monitoring wells whose screen was 4-10 m bgs.

**Table 1** Concentration in the groundwater and the isotope signature of TCE

| ID | TCE (mg/L) | cis-DCE (mg/L) | $\delta^{13}C$ of TCE (‰) | $\delta^{37}Cl$ of TCE (‰) |
|---|---|---|---|---|
| R00450 | 2.21 | 0.12 | -30.30 | -0.8 |
| R00451 | 0.73 | 0.01 | -29.43 | 0.7 |
| R00452 | <0.01 | 0.54 | -28.10 | - |
| R00612 | 0.12 | <0.01 | -27.96 | 1.3 |
| R00613 | 0.05 | <0.01 | -28.09 | 0.9 |
| SRF001 | ND | ND | - | - |
| SRF002 | ND | ND | - | - |
| SRF003 | ND | ND | - | - |
| SRF004 | ND | ND | - | - |

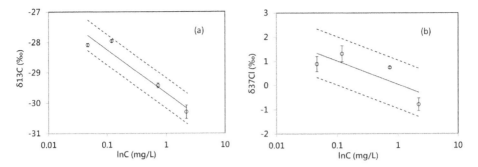

**Figure 2** *(a) $\delta^{13}C$ and (b) $\delta^{37}Cl$ values of TCE as a function of the TCE concentration in the groundwater with the error bars showing the standard deviations of triplicates and the dash lines showing the analytical tolerance*

The specific range of the depth where the source resides was unknown. When the direct evidence is desired, a high resolution characterization tool is necessary so as to acquire the depth of concern. Therefore, MIP with real-time detectors was used for this purpose.

Two transects along which MIP was conducted were depicted in Figure 3. Transect I was purposely positioned along the east boundary of the site. Since the groundwater flows primarily towards the east, the signals along Transect I can serve as a projection of the likelihood of DNAPLs.

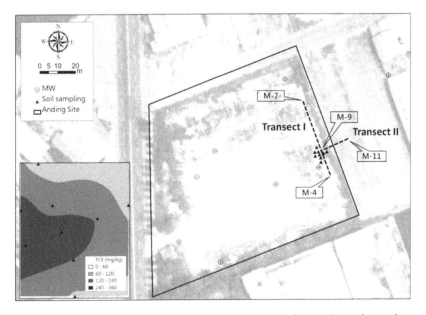

**Figure 3** *Site map showing the MIP probing transects (dash lines), the soil sampling locations (filled triangles), and the contour of TCE concentration of the soil samples at approximately 8 m bgs (the insert map)*

The ECD and DELCD results are consistent with each other showing the same locations of high response and only the ECD results were presented here. Figure 4 (a) shows the mapping of MIP-ECD signals along Transect I. The extent of darkness reflects the amplitude of ECD signals. The asymmetry of the projection is consistent with the fact that the groundwater flow is not perfectly eastbound but slightly to the north, i.e. towards M-7.

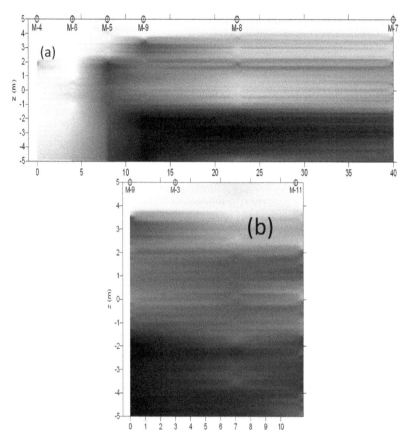

**Figure 4** *Mapping of MIP-ECD signals along (a) Transect I and (b) Transect II where the lateral axis is the distance from the leftmost point in meters and the vertical one is the elevation above the mean sea level*

The hotspot seems to reside at approximately 8 m bgs around point M-9, which is only several meters north of R00452. The relatively high response at M-9 compared to M-7 suggested that the TCE concentration at M-9 is very likely higher than that of R00450 which measured 2.21 mg/L. This indicated that the samples obtained from these locations were close to the primary source of contamination. The low concentration of TCE at the nearby well, R00452, may be explained by the phenomenon that the up-gradient concentration of a source often changes more dramatically than its down-gradient.[11] Realizing the hotspot is close to the boundary and may involve a different responsible party, it is necessary to clarify whether the source resides across the boundary or even outside the site. The mapping of transect II as shown in Figure 4 (b) clearly confirmed that

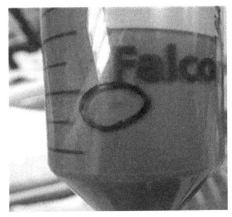 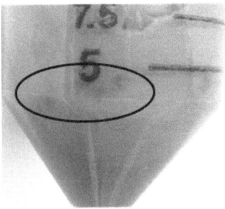

**Figure 5** *Soil samples stained by Sudan IV*

inside the site the ECD responded stronger and therefore should be the focus of the evidence collection.

Subsequently, nine soil cores were collected using Geoprobe around point M-9 as shown in the insert map of Figure 3. The depth of the soil samples to be analyzed was screened on site with the assistance of a handheld photoionization detector (PID). Screening result was consistent with the MIP-ECD signals in that both showed a strong response around 8 m bgs. Analysis results of the soil extract showed that TCE was the major component with the concentration up to 373 mg/kg. Other constituents, such as cis-DCE and 1,2-dichloroethane, were identified but in concentrations at least three orders of magnitude lower.

Despite the TCE concentration exceeding its regulated level (60 mg/kg), it is not high enough to conclude the presence of DNAPL.[7] As the residual DNAPL is likely present as ganglia, it is easy to miss the hot spot since only a small amount of soil, often less than 5 g, was sampled for analysis. Alternatively, the dye test utilizes a large amount of soil and thus increases chances of observing the DNAPL. As can be seen in Figure 5 where the soil samples were mixed with Sudan IV and some water, the circled positions where revealed a strong contrast in this black-and-white picture were dye red in reality. This is a solid evidence of showing the existence of TCE solvent, i.e., the direct evidence of the source. It may be noted that the stained portion was not significant. In fact, without carefully inspection while stirring the slurry mildly, the stain could be overlooked easily. This dye test result provides a better understanding of how the residual DNAPL could sparsely distribute and difficult to find, in which situation how and what kind of signs to look for.

## 6 CONCLUSION

Identification of pollution sources or the cause of contamination is the first and basic objective in environmental forensics. The observed pollution phenomenon is often the effect instead of the cause, but with proper tools these effects can direct us to find the evidence of the cause. In this study, we showed that isotope signature can be used to identify the number of the pollution sources and there was only one onsite source for these monitoring wells. Various tools were used to provide evidence showing that DNAPL was present at the sites close to the bottom of the sandy loam. As DNAPL was not spotted near the top of the sandy loam, the DNAPL migration pathway was not clear. However, in the old days, liquid waste might be disposed by injecting into the subsurface. Thus, such a pathway from the surface to the subsurface might not exist at this site. Although the DNAPL migration pathway was not obtained, the evidence collected here was used by the government to list the site successfully. This case study also serves as a successful field case of evidence collection.

**References**

1. D. Goldman, J. Gabry, R. Weissenborn and W. Doctor, in *Joint Services Environmental Management Conference* Former Naval Air Station Moffett Field, 2006.
2. Y. Wang, Remediation, 2013, **Spring**, 111-120.
3. Y. Wang, J. L. Newton and A. Igoe, in *Environmental Forensics: Proceedings of the 2011 INEF Conference*, The Royal Society of Chemistry, 2011, pp. 64-76.
4. R. D. Morrison and B. L. Murphy, in *Introduction to Environmental Forensics* (3$^{rd}$ edition), Academic Press, 2015, pp. 311-345.
5. S. Cretnik, A. Bernstein, O. Shouakar-Stash, F. Löffler and M. Elsner, *Molecules*, 2014, **19**, 6450-6473.
6. M. Elsner, L. Zwank, D. Hunkeler and R. P. Schwarzenbach, *Environmental Science & Technology*, 2005, **39**, 6896-6916.
7. R. M. Cohen and J. W. Mercer, *DNAPL Site Evaluation* EPA/600/R-93/022, U.S.EPA, 1993.
8. B. Jin, C. Laskov, M. Rolle and S. B. Haderlein, *Environmental Science & Technology*, 2011, **45**, 5279-5286.
9. A. Bernstein, O. Shouakar-Stash, K. Ebert, C. Laskov, D. Hunkeler, S. Jeannottat, K. Sakaguchi-Söder, J. Laaks, M. A. Jochmann, S. Cretnik, J. Jager, S. B. Haderlein, T. C. Schmidt, R. Aravena and M. Elsner, *Analytical Chemistry*, 2011, **83**, 7624-7634.
10. D. Hunkeler, N. Chollet, X. Pittet, R. Aravena, J. A. Cherry and B. L. Parker, *Journal of Contaminant Hydrology*, 2004, **74**, 265-282.
11. M. A. Guilbeault, B. L. Parker and J. A. Cherry, *Ground Water*, 2005, **43(1)**, 70-86.

# FORENSIC INVESTIGATIONS IN COMPLEX POLLUTION CASES INVOLVING PCBS, DIOXINS AND FURANS: POTENTIAL PITFALLS AND TIPS

Jean Christophe Balouet[1], Francis Gallion[2], Jacques Martelain[3], David Megson[4,5], Gwen O'Sullivan[6]

[1] Environnement International, 2 rue du Hamet, 60129 Orrouy, France
[2] Legis Environnement, BP 2006, 79403 Saint Maixent l'Ecole, France
[3] TERRAQUAtron, 61route de Saint Romain, 69450 St Cyr au Mont D'Or, France
[4] Ryerson University, 350 Victoria Street, Toronto, ON, M5B 2K3, Canada
[5] Ontario Ministry of the Environment and Climate Change, 125 Resources Road, Toronto, ON, M9P 3V6 Canada
[6] Department of Environmental Science, Mount Royal University, 4825 Mount Royal Gate SW, Calgary, AB, T3E 6K6, Canada

## 1 INTRODUCTION

Polychlorinated biphenyls (PCBs), dioxins (PCDDs) and furans (PCDFs) are frequently detected in grass, soil and farm animals near incineration plants, electric transformers recycling facilities, or after major fires during environmental forensic investigations. The sale of farm animals for food above established PCBs and PCDD/F limits is forbidden, and therefore are slaughtered, or removed from contaminated areas to graze on uncontaminated food, in an attempt to detoxify. This policy can result in substantial financial losses to farmers and other parties on whose property these contaminants are detected. If regulated levels are exceeded in one sample, the impacts are rarely limited to one area, or within the sampling period. It is therefore important to establish the extent of the contamination, distinguish the potential source(s) and identify the duration of the contamination event. These determinations are especially challenging to account for background noise or when multiple sources of these contaminants are present within a 10 kilometre (km) radius of the investigated area and could have contributed to the contamination.

This manuscript examine potential pitfalls in the analysis of (PCBs), dioxins (PCDDs) and furans (PCDFs) in terms of their regulated levels, sampling challenges, the analysis of samples and statistical interpretation of the test data.

## 2 REGULATED LEVELS

PCDD/F and PCBs are detected in the environment as complex mixtures. To assess the risks to human health and the environment, internationally agreed Toxic Equivalency Factors (TEFs) have been developed and are regularly updated by the World Health Organisation (WHO). The current TEFs are based on the findings of Van den Berg et al.[1] The biological effects of PCDD, PCDF and PCBs are mediated through the aryl

**Table 1** 2005 World Health Organization (WHO) Toxicity Equivalent Factors(TEFs)

| Dioxins | WHO TEF | Furans | WHO TEF | PCBs (Cl substitution) | WHO TEF |
|---|---|---|---|---|---|
| 2,3,7,8-TCDD | 1 | 2,3,7,8-TCDF | 0.1 | PCB-77 (34-3'4') | 0.0001 |
| 1,2,3,7,8-PeCDD | 1 | 1,2,3,7,8-PeCDF | 0.03 | PCB-81 (345-4') | 0.0003 |
| 1,2,3,4,7,8-HxCDD | 0.1 | 2,3,4,7,8-PeCDF | 0.3 | PCB-105 (234-3'4') | 0.00003 |
| 1,2,3,6,7,8--HxCDD | 0.1 | 1,2,3,4,7,8-HxCDF | 0.1 | PCB-114 (2345-4') | 0.00003 |
| 1,2,3,7,8,9-HxCDD | 0.1 | 1,2,3,6,7,8-HxCDF | 0.1 | PCB-118 (245-3'4') | 0.00003 |
| 1,2,3,4,6,7,8-HpCDD | 0.01 | 1,2,3,7,8,9-HxCDF | 0.1 | PCB-123 (345-2'4') | 0.00003 |
| OCDD | 0.0003 | 2,3,4,6,7,8-HxTCDF | 0.1 | PCB-126 (345-3'4') | 0.1 |
| | | 1,2,3,4,6,7,8-HpCDF | 0.01 | PCB-156 (2345-3'4') | 0.00003 |
| | | 1,2,3,4,7,8,9-HpCDF | 0.01 | PCB-157 (234-3'4'5') | 0.00003 |
| | | OCDD | 0.0003 | PCB-167 (245-3'4'5') | 0.00003 |
| | | | | PCB-169 (345-3'4'5') | 0.03 |
| | | | | PCB-189 (2345-3'4'5') | 0.00003 |

**Table 2** WHO established Toxicity Equivalent (TEQs) for meat and grass, 1998 and 2005 values in picogram per gram (pg g$^{-1}$) and nannogram per gram (ng g$^{-1}$)

| Meat FAT | Toxicity Factor | TEQ Σ(PCDD+PCDF+PCBDL) pg g$^{-1}$ I TEQ WHO | PCB ind ng g$^{-1}$ |
|---|---|---|---|
| 2011 | I-TEF WHO 1998 | 4,5 pg g$^{-1}$ I-TEQ WHO 1998 | Not applicable |
| 2012-13-14 | I-TEF WHO 2005 | 4 pg g$^{-1}$ I-TEQ WHO 2005 | 40 ng g$^{-1}$ |
| Grass | Toxicity Factor | TEQ Σ(PCDD+PCDF+PCBDL) pg g$^{-1}$ I TEQ WHO | |
| 2011 | I-TEF WHO 1998 | 1,25 pg g$^{-1}$ I-TEQ WHO 1998 | |
| 2012 | I-TEF WHO 2005 | 1,25 pg g$^{-1}$ I-TEQ WHO 2005 | |

hydrocarbon receptor (AhR) which has a high affinity for 2,3,7,8-substituted PCDD/Fs and co-planar PCBs with either one or no chlorines in the 2, 2', 6 or 6' positions. Table 1 lists the 17 PCDD/Fs and 12 PCBs with TEFs from Van den Burg et al.[1]

Threshold and alert levels have been established by WHO for different matrices, and are expressed in TEQ (Toxicity Equivalent), for PCB DL (dioxin like), dioxins (PCDD) and furans (PCDF). More recently, other threshold levels have also been established for the 7 indicator PCBs (PCBIs) (PCB-28, PCB-52, PCB-101, PCB-118, PCB-138, PCB-153, PCB-180). Common limit values can be established as follows for samples such as cattle and grass based on WHO TEFs (Table 2). Alert levels have also been generated for grass (12% humidity) at 0,5 pg g$^{-1}$ (picogram per gram)WHO -PCDD/F-TEQ and 0,35 pg g$^{-1}$ WHO-PCBDL-TEQ. (e. g. European Directive 2006/13/CE)

## 3 SAMPLING

Numerous samples are usually gathered during the course of an environmental forensics investigation. These might be collected from air particles (on an Owen gage and from emission source), as well as gas phases, vegetation, soils and animals, leading to complex data set with the tens of thousands of data points.

There are several pitfalls when selecting appropriate samples, including the following:

- When sampling air and particulate matter to consider and monitor wind direction. The sampling strategy should be designed to include a background sample which is upwind of the suspected source(s). Most passive samplers collect dust from all directions and so need to be combined with meteorological data to establish sources. However there have been recent advances in sample collection technologies that allow 360° sampling at 15° intervals to establish sources of dust contamination.[2]

- When sampling the grass, it is essential to understand that these contaminants will be found at higher concentration in winter, during the plant's dormant season, and at lower concentrations during spring due to grass growth. Other complex mechanisms such as microbial activity, photolysis and selective evaporation are identified, which can reduce concentrations or alter chemical signature / fingerprint (i.e. less chlorinated molecules are more degraded than the more chlorinated ones).

- When sampling surface soils the samples are all made at same depth (between 0 and 5 cm maximum) as deeper soils are likely to be less contaminated. Similarly, it is important to check that soils have not been perturbed. i.e. ploughed, or disturbed by cattle feet in muddy areas, which would result in contaminants' dilution.

- When sampling animals if the goal is to compare total concentrations then it is preferable to obtain samples from the same tissue type as concentrations vary in different tissues based on the lipid content. Therefore all data should be lipid corrected to allow for a representative comparison between samples. Concentrations can vary in different tissue types however the signature appears to be very consistent, indicating that whilst collection of the same tissue type is preferable, the signature in different animals can be compared using samples obtained from different tissue types.[3]

- Impact to farmland, crops and livestock also depend on distance, wind direction, topography. In some very rare instances, when farm animals are sampled from freezers, expert needs to check that selected meat does not originate from distant areas.

- The background concentrations and signature need to be established during investigation and several samples are therefore required from several distant areas, away from source, up to 10 km (see Figure 3 for example).

In all cases, samples should be properly labelled and stored in containers appropriate for the analysis that is being undertaken. As some PCB, PCDD/F congeners are semi volatile samples should be packed tightly in glass containers to reduced headspace and stored in cool boxes at 4°C once collected, to reduced potential losses through evaporation. All samples should be transported to the laboratory as soon as possible (within 24 hrs) with appropriate chain of custody documentation.

# 4 ANALYSES AND STATISTICAL INTERPRETATION

## 4.1. Analysis

When regulatory authorities are involved, 2,3,7,8 substituted PCDD/F and WHO12 PCBs are often of forensic interest (Table 1). For these contaminants, GC-MS or GC-ECD provide a mean to quickly determine whether a sample is grossly contaminated. Many commercial laboratories provide limits of detection (LODs) in the range of 10 – 1000 µg kg$^{-1}$. This can be a significant limitation as background total PCB concentrations (in UK urban soils) are between 0.01 – 40 µg kg$^{-1}$.[4] Therefore, sample clean-up and analysis by HRMS is often required to improve limits of detection to less than 1 µg kg$^{-1}$ by removing or filtering out many interfering compounds. Whilst the 2,3,7,8 substituted PCDD/F and WHO12 PCBs can establish risks to human health, additional congeners may be needed to identify the source. This can be a time intensive task using GCMS or GCHRMS as several runs using different column types are needed to produce a comprehensive congener specific data set, due to multiple co-elutions of PCB and PCDD/F congeners (there are over 400 PCBs and PCDD/Fs).

The development of comprehensive two-dimensional gas chromatography (GCxGC-TOFMS) provides an extra dimension of separation which significantly increases the resolving capacity. This has allowed for the identification of over 190 individual PCB congeners along with simultaneous identification of other organohalogenated contaminants.[5,6] Figure 1 displays the separation of 173 peaks from 209 PCBs achieved using a PCB specific Rtx-PCB column on GC-MS, compared to 200 peaks from 209 PCBs which was achieved using GCxGC-TOFMS (Rtx PCB and Rxi-17 columns). However, such expertise is usually at great costs, for example analytical costs for the PCBDL, PCDD / PCDF and PCBIs range in the 500 US$, and go as high as 2000 $ for documenting all PCB and PCDDF congeners.

## 4.2. Statistical Interpretation

In most investigations involving PCBs, PCDD/Fs the results are screened against a specific health value to establish the risks. This can be a simple comparison of individual sample concentrations against a threshold level, or the use of statistics to determine if an area (or site) is above the threshold value at a 95% degree confidence level. However, for many forensic investigations the goal is not to determine if the concentration is above a threshold, but to establish where the contamination has originated from. In these instances, absolute concentrations are often of little use and signatures based on the relative proportions of individual congeners are more useful, especially when combined with multivariate statistics.

Statistical methods are of major relevance when comparing signatures through fingerprinting: i.e. using R2 to determine the confidence interval, and multivariate analysis to compare hundreds of samples and dozens of compounds. Most statistical techniques require some prior knowledge of the contamination event. However, some systems with multiple sources and degradation pathways are too complex to establish a priori assumptions;[7] in such investigations, exploratory data analysis techniques, such as principal component analysis and polytopic vector analysis, can be used to identify the source(s) of contamination. Principal component analysis may be especially useful for identifying temporal patterns in the data. In some cases it can be important to identify if a change in the signature has occurred after a specific date (for example when a company was supposed to implement procedures to reduce its emissions to air). Figure 2 shows how principal component analysis was used to show that the absence of a distinguishing pattern in dust collected over a three year period.

In these cases it is also important to encompass and distinguish other potential contributors such as domestic heating, backyard burning, slash and burn as well as the background noise: The US EPA have produced a list of potential contributors, which is useful for forensic investigations, and is accessible at http://www.epa.gov/ncea/pdfs/dioxin/2k-update/. Historical research and proper investigations are essential to characterize potential sources so as to distinguish liabilities and allocate torts as honestly as can be. It is here important to characterize the background noise: when threshold limit values are exceeded in grass and cattle, the background noise typically accounts for around 2 % of the contamination.

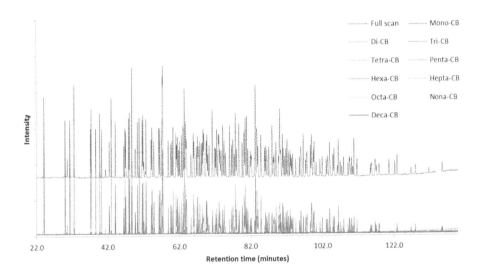

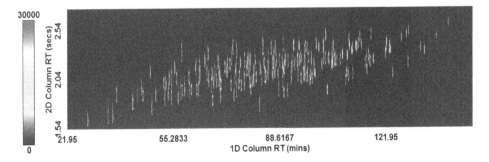

**Figure 1** *Comparison of PCB separation using GC-MS (above), compared to GCxGC-TOFMS (below). Figure adapted from Megson et al. 2014.*[3]

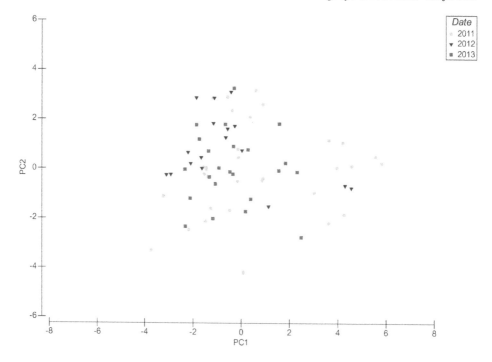

**Figure 2** *PCA scores plot showing no noticeable change in dust signature over a three year monitoring period*

### 4.3 Modelling Data

Expertise is required to identify the source by mapping the plume impacts for the different environmental media and receptors and comparing the chemical fingerprint of contaminants sampled at the source and at a distance. The concentration maps can be centred by source, when the impact distance can exceed 6 km, as documented for PM10, [8] depending on wind conditions or height of emission source.

Documenting the exact contribution of suspected source versus other potential contributors is difficult. Some simple methods allow for the calculation of how much the supposed source has been contributing, within a given distance to source, using the following parameters and formulae:

- Annual emissions: $Em_{ann}$;

- Distance to source: $Dis_{source}$ used as a radius. Several models exist to estimate particle concentrations, according to their size, wind speed, and height of chimney as release point;

- Annual mass production for grass by surface –dry weight = 12% Humidity: $Prod_{mass}$ in kg/m$^2$ or ton/hectar (1000 kg/10 000m$^2$), for example 6 tons of hay per hectare;

- Concentration: Such a simulation can be operated on regulated level or on observed concentrations $Conc_{reg}$ or $Conc_{obs}$ or to estimate impacts at a given level $Conc_{theor}$.

Concentrations are usually expressed in pg g$^{-1}$ TEQ for PCBDLs, PCDDs, PCDFs, and in ng g$^{-1}$ PCBI, and

- Impacted surface: Surf$_{imp}$.

A simple formula can be based under the assumption that the contamination results in homogenous concentrations, independent on distance or winds direction or speed. Contaminant concentration in a given radius:

$$\text{Conc}_{theor} = \text{Em}_{ann} / \text{Dis}_{source}^2 \times \pi \times \text{Prod}_{mass} \qquad (1)$$

With an example of Em$_{ann}$ 0,5 g TEQ, Dis$_{source}$ 3000 m, Prod$_{mass}$ 6 tons ha$^{-1}$ or 0.6 kg per m$^2$, Conc$_{theor}$ = 0,29 pg g$^{-1}$ TEQ

The same formula can be used to determine the theoretical radius within which a regulated level would be found, if homogenous concentrations. In the preceding example, the threshold limit of 0,35 pg g$^{-1}$ TEQ PCBDL for grass is exceeded within a radius of 3.2 km. The meat fat in most cows, however, would contain TEQ PCBDL values exceeding this value whilst veals contamination is commonly twice higher than milking cows. These calculations are usefully complemented by congener fingerprint profiles calculated from samples. Figure 3 presents the following theoretical profile for vegetal concentrations from samples made within same wind direction, and at varying distances. The abscissa –in meters- represents the distances to source (upwind in negative numbers and downwind in positive numbers) and ordinates represent the pollutant concentrations.

Such simple graphs, derived from Excel, can be used to estimate concentrations Conc$_{theor}$ at Dis$_{source}$, whether upwind or downwind. Such quasi-Gaussian profile further enables the possibility to establish that:

- the higher concentrations are by source;
- impacts are slightly higher downwind;
- the background noise is found at 10 km distance, upwind and downwind, and
- the absence of secondary sources that would have caused significant / local or temporal concentration anomalies.

In this theoretical case, the background noise is too low to cause concentrations in excess of regulated thresholds, as is an almost universal condition, though depending on the possibly high concentration of industrial plants, sources that would exist nearby.

## 5 DEBATES AND POTENTIAL PITFALLS

### 5.1 Time

Time is a critical factor because it can take years before the pollution is first documented.

To the parties and their consultants, it is primordial to focus on the situation at the time the case and the supporting evidence did build, possibly years before forensic involvement. Dendrochemical evidence gained from pine or picea needles can help

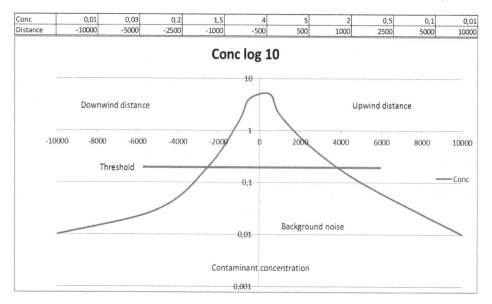

**Figure 3** *Theoretical concentration profile for vegetal samples made within same wind direction, and at varying distances.*

document current and past contamination, typically year by year and back to 7 or 8 years before sampling time, also providing the chemical fingerprint of the 35 sought congeners.

In cases where an industrial party is continuously involved, they will do their best to abate their emissions, would it be by mechanical / physical / chemical control systems, or by reducing their activities, either on an annual basis, or within weeks or months before analytical campaigns are conducted, of which they proactively are informed. In such a situation, measured concentrations would not exceed regulated thresholds, although they may have at other times.

Furthermore, environmental forensic expertise also can address preventive considerations: this *in futurum* part of experts' tasks is to check whether proper solutions are in place or additional measures should be set. Such anticipative considerations are essential to help the judge and law enforcement authorities make sure that the pollution problem will not continue after litigation of the dispute, or would not occur in other circumstances.

With the adoption of international conventions, such as the 1989 Aarhus Convention or Sustainable development, the experts' mandates are turning more and more complex as they involve considerations to Human Rights, or environmental damage, that goes far beyond common forensic practice in the 20th Century.

### 5.2 Combining Datasets

Care is required when combining and comparing datasets from different sampling investigations, as analytical procedures are constantly changing. Similar challenges exist when comparing samples from different matrixes as the extraction and clean up procedures will differ depending on the matrix and possibly even the sample. This is especially important when comparing low level concentration data, in instances where the concentration is below the limit of detection (LOD) a value of the LOD or LOD/2 is often substituted.[9] If the LOD is different between the two or more datasets this can indicate a difference between the datasets which may not actually exist in the samples.

Another consideration with PCB and PCDD/F data collected over different time periods is the fact that the TEFs are regularly reviewed and updated by WHO. The most recent values were released in 2005 and so it is important to understand that any TEQs produced before 2005 may have been calculated using the 1998 WHO TEFs and any TEQs calculated before 1998 may have been calculated using the 1988 NATO I-TEQs. Depending upon which of the three TEFs are used, and on the congener mixture in the samples, it is not uncommon for the calculated TEQs to vary by over 10%.[10] Therefore care must be taken when comparing and combining datasets over different time periods and it is always useful to double check the raw data to identify which TEFs have been used in any TEQ calculation.

**5.3 Evidence**

In some cases local events, such as slash and burn, backyard waste burning may be inferred as temporal and local cause of exceeding threshold limits in some samples. A rule of thumb is that such events also exist in the areas where background noise is documented. Some parties may utilize selected data rather than considering the entire data set in an attempt to advocate a particular position, rather than considering all of the information and basing scientific opinions on this analysis.

The bibliographic resources available are vast, technically complex, and may be available from governmental and intergovernmental organizations or from peer-reviewed publications. Language is a significant barrier. This would require considerable resource as dozens of key references are several hundred pages, when it may be sufficient to translate the important pages only (for example the web link given in section 4.2. refers to an US EPA document several hundred pages long).

Depending on the judicial system in place in the country where the case takes place, it is more or less easy to access all pre-existing evidences. Taking just one example, when many others could be referred to, let's consider that an industrial plant has been found to cause environmental impacts in excess of regulated level; worker's staff, most directly exposed to industrial activity, may have been subject to biological sampling, to document their exposure to the specific pollutant. What about if such worker's exposure has been documented by company and his exposure level is not communicated to the experts, nor attorneys and even not Court, whenever the numbers may exceed by 50 or 100 times the regulated ones? Such evidence would have proved essential to link contamination to the workplace as a source, but also to further establish toxic torts.

6 CONCLUSIONS

Environmental forensics investigations are key to Justice by honestly documenting the facts, understanding each and all of the evidences would they be served by plaintiff, defendant, or newly established as part of a litigation, whatever the data and influencing parameters can be complex. Contradictory debate is key, provided it is served with unbiased transparency and highest competent objectivity; some evidences may prove scientifically opposable, others turn as highlights when to form an honest opinion.

**Acknowledgements**

This publication is the result of collegial expertise within the French Company of Judicial Environmental Experts (Compagnie Nationale des Experts de Justice en Environnement) along with INEF.

**References**

1. M. Van den berg, L. S. Birnbaum, M. Denison, M. De Vito, W. Farland, M. Feeley, H. Fiedler, H. Hakansson, A. Hanberg, L. Haws, M. Rose, S. Safe, D. Schrenk, C. Tohyama, A. Tritscher, J. Tuomisto, M. Tysklind, N. Walker & R. E. Peterson. 2006. The 2005 World Health Organization reevaluation of human and mammalian toxic equivalency factors for dioxins and dioxin-like compounds. *Toxicological Sciences*, 93, 223-241
2. J. Bruce, H. Datson, J. Smith, M. Fowler. (2015). Characterisation and modelling of dust in a semi-arid construction environment. In G. O'Sullivan and D. Megson (eds). *Environmental Forensics: Proceedings of the 2014 INEF Conference*.
3. D. Megson. 2014. Application of Polychlorinated Biphenyl Signatures for Environmental Fingerprinting. PhD Thesis, Plymouth University.
4. C.S. Creaser, M.D. Wood, R. Alcock, D. Copplestone & P J. Crook. 2007. UK Soil and Herbage Pollutant Survey Report No. 8. Environmental concentrations of polychlorinated biphenyls (PCBs) in UK soil and herbage. Environment Agency, Science Project Number SC000027.
5. J.F.Focant, A. Sjodin, W.E.Turner, & D.G. Patterson Jr. 2004. Measurement of selected polybrominated diphenyl ethers, polybrominated and polychlorinated biphenyls, and organochlorine pesticides in human serum and milk using comprehensive two-dimensional gas chromatography isotope dilution time-of-flight mass spectrometry. *Analytical Chemistry*, 76, 6313-6320.
6. D. Megson, R.B. Kalin, P. Worsfold., C. Gauchotte-lindsay, D.G. Patterson Jr, M.C. Lohan, S. Comber, T.A. Brown & G. O'Sullivan. 2013. Fingerprinting polychlorinated biphenyls in environmental samples using comprehensive two-dimensional gas chromatography with time-of-flight mass. *Journal of Chromatography A*, 1318, 276-283.
7. G.W. Johnson, J.F. Quensen III, J.R. Chiarenzelli & C.M. Hamilton. (2006) 'Polychlorinated Biphenyls'. In Morrison, R.D. and Murphy, B.L. (eds.) *Environmental Forensics Contaminant Specific Guide*. Academic Press.
8. S. Denys, D. Gombert, K. Tack. 2012. Combined approaches to determine impact of wood fire on PCDD/F and PCB contamination of the environment: a case study. *Chemosphere* 88, 806-812.
9. CIEH (Chartered Institute of Environmental Health) and CL:AIRE (Contaminated Land: Applications in Real Environments). 2008. Guidance on Comparing Soil Contamination Data with a Critical Concentration.
10. D. Megson and S. Dack. (2011). Assessing changes to the congener profile of PCDD and PCDF during bioaccumulation in chicken and duck eggs. In R.D. Morrison, and G. O'Sullivan (eds) *Environmental Forensics: Proceedings of the 2011 INEF Conference*. 224-261.

USE OF SYMMETRIC TETRACHLOROETHANE TO AGE DATE CHLORINATED SOLVENT RELEASES

Robert D. Morrison

Independent Consultant. 56-2773 Lahuiki Place, Hawi, HI. 96719, USA

1. INTRODUCTION

Manufacturing impurities in chlorinated solvent products are of significant forensic importance for bracketing when a release occurred and/or identifying the source. While forensic techniques traditionally used for age dating and source discrimination, such as compound specific isotope analysis (CSIA), isotopic and chemical ratio analysis, reconstructed plume techniques, surrogate chemical indicators and stabilizers are available, less emphasis has been given to the use of manufacturing impurities as a corroborative technique.[1]

While first reported in 2009 by Morrison and Hone at the INEF Conference at St. John's College in Cambridge and further refined as a forensic technique in 2013, the use of manufacturing impurities in PCE (tetrachloroethylene) and TCE (trichloroethylene) produced via the oxychlorination of acetylene or ethylene offers a means to estimate when these chlorinated solvents were produced.[2,3] An assumption with this technique is that manufacturing impurities associated with acetylene as a PCE and TCE feedstock ceased in 1978 in the U.S. when it was replaced by ethylene.[2] Of the various manufacturing impurities reported with TCE, PCE and methyl chloroform, the detection of 1,1,2,2-tetrachloroethane (1,1,2,2-TeCA (acetylene tetrachloride in the early literature) provides the means to identify that the solvent was manufactured prior to about 1977.[4] Of forensic interest is that in 1977, methyl chloroform and 1,2-DCA (1,2-dichloroethane) were cited as likely TCE impurities when produced from ethylene but not TeCA. TCE and PCE is also produced via the direct thermal chlorination of methane, ethane, propane, propylene or their partially chlorinated derivatives.[5,6,7,8,9]

## 2. HISTORY, PHYSIOCHEMICAL PROPERTIES AND PRODUCTION OF TeCA

### 2.1 Historical Use of TeCA

TeCA was first synthetized in 1869 with large scale production beginning in 1903 and continuing through World War I (WWI). During WWI, TeCA was used in the varnish sprayed on linen airplane wing surfaces, due in part to its unique ability to tighten the stretched fabric over the wings. By the end of WWI numerous cases of poisonings were attributed to TeCA in the aircraft industries of Germany, France, England and Holland; in England alone, there were over 70 reported cases with 12 deaths. TeCA was subsequently replaced by solvents believed to be less toxic.[10]

In World War II, the United States (U.S.) Army used TeCA to impregnate clothing for protection against mustard gas and for spotting in fabric cleaning, the artificial silk industry and for manufacturing artificial pearls.[11] A 1946 article lists impurities associated with TeCA as including 1,2-dichloroethane (0-2%), pentachloroethane (0-8%) and hexachloroethane (0-4%).4

In addition to its use as a solvent for cellulose acetate, waxes, greases, rubber, and sulfur,[12] TeCA was a major component of a decontaminating agent produced by the U.S. military. Other uses include its use as an organic solvent process to manufacture chemical-agent-resistant clothing, an extraction solvent in the pharmaceutical industry, metal degreasing, as an ingredient in paint removers, varnishes, lacquers, photographic film, rust removers, resins and waxes,[13] as an oil and fat extractant,[14] as an alcohol denaturant; as an ingredient used in the synthesis of cyanogen chloride, polymers, tetrachloro-alkylphenol and as a solvent in the preparation of adhesives,[15] to determine the theobromine content in cacao, as a solvent to clean chemical weapon components,[16] as an immersion fluid in crystallography, used in biological laboratories to produce pathological changes in the gastrointestinal tract, liver and kidneys, used to estimate the water content of tobacco and many drugs, as a solvent for impregnation of furs with chromium chloride and an ingredient in pesticides,[17] although it is not currently registered in the U.S. Prior to 1967, TeCA was the primary starting material for producing chlorinated hydrocarbon solvents but since that time, the use of ethylene has predominated.

### 2.2 Physiochemical Properties

TeCA is a synthetic chemical with no natural sources.[17] Given that TeCA is infrequently studied or detected relative to the more prominent chlorinated solvents, a brief summary of its physical and chemical properties is warranted (Table 1).3,[18,19,20,21]

A review of the literature regarding TeCA degradation is consistent with the production of the major (*cis* and/or *trans*-1,2-DCE TCE and 1,1,2-TCA) products as well as others depicted in Figure 1.[16,22,23,24,25]. TeCA degrades under anaerobic conditions by (1) hydrogenolysis that produces 1,1,2-trichloroethane (1,1,2-TCA) and 1,2-dichloroethane (1,2-DCA) as intermediate daughter compounds, (2) dichloroelimination ($C_2H_aCl_b + [2H] \rightarrow C_2H_aCl_{b-2} + 2HCl$) that produces 1,2-dichloroethene (*cis* and *trans* 1,2-dichloroethylene) and vinyl chloride as intermediate daughter products and (3) via abiotic dehydrochlorination ($C_2H_aCl_b + [2H] \rightarrow C_2H_{a+1}Cl_{b-1} + HCl$) that produces TCE (Figure 1).[26,27,28,29,30,31] Chen et al., 1996 reported that all three reactions occur simultaneously with TeCA under methanogenic

## Use of Symmetric Tetrachloroethane to Age Date Chlorinated Solvent Releases

**Table 1** *Physical and chemical properties of 1,1,2,2-tetrachloroethane*

| Properties | Values |
|---|---|
| Molecular Weight | 167.8 |
| Molecular Formula | $C_2H_2Cl_4$ |
| Melting Point (°C) | -42.4 |
| Density (g/cm$^3$) | 1.6 |
| Boiling Point (°C) | 146.2 |
| Solubility in Water at 20°C (mg/l) | 2,870 |
| Absolute Viscosity (cP) | 1.76 |
| Kinematic Viscosity (cS) | 1.12 |
| Vapor Pressure at 25°C (mm Hg) | 6.3 |
| Henry's Law Constant (atm-m$^3$/mol) | 3.56-4.55x10$^{-4}$ |
| Octanol Water Partition Coefficient | 2.56 |
| Relative Vapor Pressure | 5.79 (air = 1) |
| Specific Gravity at 20°C | 1.596 |
| Relative Vapor Density | 1.04 |

conditions using municipal digester sludge; dehydrochlorination and dichloroelimination were cited as more important in the initial degradation of TeCA and 1,1,2-TCA.[28] 1,2-DCA and other intermediates are mainly degraded by reductive dechlorination.[28]

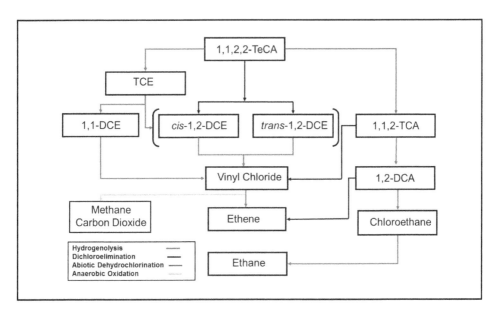

**Figure 1** *1,1,2,2-tetrachloroethane degradation pathways*

## 2.3 Production of TeCA

TeCA was last produced as an end product [a] in the U.S. by the Specialty Materials Division of Eagle-Pircher Industries in Lenexa, Kansas. By the late 1980s, the Eagle-Pircher facility was sold to the Vulcan Materials Company and TeCA production was discontinued.[32] This chronology parallels production patterns in Canada, where the last plant in Shawinigan, Quebec to manufacture TeCA as an intermediate for the production of TCE and PCE [b] ceased operations in early 1985. [33,34] Other literature cites TeCA production in the U.S. and Canada as an end product ending in the early 1990s.[15,35]

## 3. TeCA AS A MANUFACTURING IMPURITY IN CHLORINATED SOLVENTS

### 3.1 Production of PCE and TCE from Acetylene

PCE and TCE were historically produced in an integrated manufacturing process. For PCE using acetylene and chlorine as the primary raw materials, the acetylene is chlorinated with ferric chloride and sometimes phosphorus chloride and antimony chloride as catalysts.2 For the production of TCE, the TeCA is dehydrochlorinated to form TCE often in aqueous bases, such as calcium hydroxide or via thermal cracking with barium chloride on activated carbon or silica or aluminum gels. The TCE is then chlorinated to form pentachloroethane that is dehydrochlorinated to produce PCE as described by the following reactions.[36]

$$CH \equiv CH \ (acetylene) + 2Cl_2 \rightarrow CHCl_2CHCl_2 \ (1,1,2,2\text{-}tetrachloroethane) \tag{1}$$

$$CHCl_2CHCl_2 \rightarrow CHCl=CCl_2 \ (trichloroethylene) + HCl \tag{2}$$

$$CHCl=CCl_2 + Cl_2 \rightarrow CHCl_2CCl_3 \ (pentachloroethane) \tag{3}$$

$$CHCl_2CCl_3 \rightarrow CCl_2 = CCl_2 \ (perchloroethylene) + HCl \tag{4}$$

While the catalytic oxidation of TeCA proceeds, other reactions produce impurities in the PCE/TCE product streams, including cis and trans-1,2-dichloroethane, carbon tetrachloride (a co-product), 1,1,2-trichlorethane, hexachloroethane, hexachlorobenzene and pentachloroethane. [37,38,39,40,41] In the environment, TeCA also degrades abiotically to TCE (Figure 1) and thence to 1,1-DCE, cis/trans-1,2-DCE, vinyl chloride and ethane.[42] CSIA may provide a means to confirm whether TeCA, TCE and its daughter products originated with its production or whether TCE and its degradation products are associated with another source.

Prior to 1950, TCE was produced almost exclusively from acetylene and accounted for 85% of total U.S. production between 1963 and 1967 (the remaining 15% from ethylene).2 By the early to mid-1970s, about 8% of the TCE produced used acetylene as production transitioned from acetylene to the less expensive ethylene, supposedly resulting in a purer

---

[a] Trade names include Acetosol, Bonoform, Boroform and Cellon (E. Bingham, B. Cohrssen, C. H. Powell in Patty's *Toxicology* 5th ed, Vol. 5. John Wiley & Sons. New York, NY, 2001, 36).
[b] TCE and PCE have not been manufactured in Canada, since 1986 and 1992, respectively.

product.[43,44] Of the five facilities[c] in the U.S. producing 217.6 10³ metric tons of TCE in 1975, only the Electrochemical Division of Hooker Chemicals and Plastics Corporation in Taft, Louisiana used acetylene with the other four facilities using ethylene as a feedstock.[45,46,47] In January of 1978, Hooker Chemicals closed the plant in Taft.[48,49]

The earliest commercial process used acetylene to produce PCE.[50] PCE production using TeCA was described in a 1944 U.S. patent describing the reaction of oxygen with symmetrical TeCA at a temperature of 300° to 600°C in the presence of a copper oxide catalyst.[37] A 1959 U.S. patent describes a reaction in which 0.4 to 0.6 mole of pure oxygen per mole of TeCA was reacted with a catalyst consisting of an inert porous carrier impregnated with cupric and zinc chloride containing from 5 to 45% by weight at a temperature between 298 and 499°C.[39] Exceptionally high yields of PCE were obtained with this cupric-zinc chloride catalyst. By 1970 most PCE in the U.S. was produced by the thermal chlorination of propane. In 1972, after two major acetylene-based plants in the U.S. were closed, only about 5% of the PCE production used acetylene. In 1975, the bulk of PCE output (97%) was derived from the oxychlorination of ethylene dichloride or by the simultaneous chlorination and pyrolysis of hydrocarbons such as propane. A minor amount of PCE (about 3%) was produced from acetylene.[51] In January of 1978, the last plant using acetylene for PCE production as a feedstock in Taft, Louisiana (Hooker Chemicals and Plastics Corporation) ended.[41,46,52]

## 3.2 Production of PCE and TCE from Ethylene

In the ethylene process, ethylene is chlorinated to form 1,2-dichloroethane (1,2-DCA) which since 1972, was then chlorinated to yield 1,1,1,2-tetrachloroethane followed by the elimination of hydrochloric acid to yield TCE.[8,53] In 1975, about 90% of the U.S. commercial production of TCE used ethylene as a feedstock.1

PCE is co-produced with TCE from 1,2-DCA. Product ratios are varied from nearly 100% PCE by adjusting the feedstock proportions and process conditions. The principal reactions in this process are:

$$2C_2H_4Cl_2 + 5Cl_2 \rightarrow C_2H_2Cl_4 + C_2HCl_5 + 5HCl \tag{5}$$

$$C_2H_2Cl_4 + C_2HCl_5 \rightarrow C_2HCl_3 + 2HCl + C_2Cl_4 \tag{6}$$

$$7HCl + 1.75O_2 \rightarrow 3.5 H_2O + 3.5 Cl_2 \text{ (Deacon Process)}^d \tag{7}$$

$$2C_2H_4Cl2 + 1.5Cl_2 + 1.75O_2 \rightarrow C_2HCl_3 + C_2Cl_4 + 3.5H_2O \tag{8}$$

---

[c] The five U.S. facilities producing TCE in the U.S. were the Diamond Shamrock Chemical Co., in Deer Park, TX, Dow Chemical, in Freeport, TX, Ethyl Corporation in Baton Rouge, LA, Hooker Chemical in Taft, LA and PPG Industries in Lake Charles, LA. (U.S. Environmental Protection Agency, Trichloroethylene Status Assessment of Toxic Chemicals, EPA-600/2-2-79-210m, Cincinnati, OH, 1979, 7).

[d] A method of chlorine production by passing a hot mixture of gaseous hydrochloric acid with oxygen over a cuprous chloride catalyst.

## 3.3 Production of Methyl Chloroform and Vinyl Chloride Monomer

The manufacture of methyl chloroform and vinyl chloride monomer (VCM) can produce minor quantities of TeCA as a by-product, although most of it is recycled for the production of other chlorinated solvents or incinerated.[48] Figure 2 is a conceptual flow diagram of an integrated chlorinated manufacturing process depicting the TeCA production as an intermediary in the formulation of methyl chloroform.[54]

The historical production of methyl chloroform includes its derivation from vinyl chloride, vinylidene chloride (1,1-dichloroethane) and ethane. In 1976, 63% of the U.S. methyl chloroform production was from vinyl chloride (Dow, Freeport, TX), 28% from ethane (Vulcan Materials Co., Geismar, LA) and 9% from vinylidene chloride (PPG Industries, Inc., Lake Charles, LA). Of these three feedstock, vinyl chloride is the most common.

Although these processes can produce TeCA as an intermediary or as a by-product, it has not been identified as a manufacturing impurity in methyl chloroform in the U.S..[55,56,57,58,59,60,61,62] A report by Environment Canada cited the presence of small amounts of TeCA in waste materials from the manufacture of methyl chloroform although estimates of the quantities are no longer available (as of June 1992, methyl chloroform is no longer manufactured in Canada.[33] Some TeCA, however, has been reported in the tars and heavy ends associated with the production of VCM.[63] Gruber 1976 reports VCM heavy ends as containing TeCA (38%) from a distillation still or about 0.014 kilogram per kilogram of VCM product.[64]

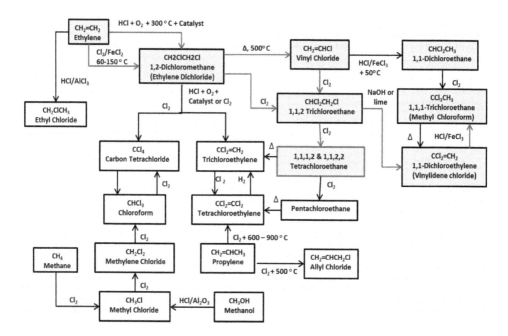

**Figure 2** *Hypothetical integrated manufacturing process for the production of chlorinated hydrocarbons with TeCA as an intermediate product*

## 4. TECA AS A DIAGNOSTIC INDICATOR

Absent other information, the detection of TeCA with TCE, PCE and methyl chloroform in an environmental sample is indicative of its production prior to 1978, while its presence without chlorinated solvents or its release as a discrete source as indicative of its manufacture prior to about 1985. When present with stabilizers (e.g., 1,4-dioxane, epichlorohydrin, etc.) with known production histories, TeCA provides another chemical opportunity to bracket when these chlorinated solvents were produced.[65] Additional opportunities include identifying known stabilizer associations such as 4-methylmorpholine, cyclohexane oxide in TCE, PCE and 1,4-dioxane, *n*-methyl-1-methanamine or formaldehyde dimethyl hydrazine with methyl chloroform to discriminate between a release of a chlorinated product containing TeCA from a discrete release of TeCA. Other useful associations include the presence of acetylene based manufacturing impurities other than TeCA in a sample with TCE and PCE, including 1,1,1,2 and 1,1,2-TCA7.[66,67,68] and an association of 1,2-DCA with TCE produced using ethylene as the feedstock, indicating a post 1978 manufacturing impurity [e].[70]

## 5. CONCLUSION

As with any forensic investigation, the use of corroborative information is useful for confirming the association of TeCA with a chlorinated product produced prior to 1978. The use of multiple diagnostic indicators is especially useful when examining TeCA as it may be present as an on-site process aid or as a formulation component, as contrasted with its use with the production of a chlorinated solvent or as an end product.[18,69]

When using PCE and/or TCE manufacturing impurities in an environmental investigation, care is required, to identify whether (1) the chemical is a manufacturing impurity, (2) an intentional ingredient in the product or (3) originated as a discrete release that co-mingled with PCE and/or TCE. Many other stabilizer and manufacturing opportunities are available in the scientific literature to use with TeCA to bracket when the PCE, TCE and/or methyl chloroform was produced.[2,3]

## References

1  R. D. Morrison and B. L. Murphy, 2015 in *Environmental Forensics 3<sup>rd</sup> Ed.*, ed. B. L. Murphy and R. D. Morrison. Academic Press, Oxford, UK, 2015, 312-335.

2  R. D. Morrison and J. Hone, *Environmental Forensics Proceedings of the 2009 INEF Calgary Conference* ed. R. D. Morrison and G. O'Sullivan, Royal Society of Chemistry, Oxford, UK. 2010, 289-304.

3  R. D. Morrison and B. L. Murphy, in *Chlorinated Solvents A Forensic Evaluation* ed. R. D. Morrison and B. L. Murphy Royal Society of Chemistry, Cambridge, UK, 2013, 60-65.

4  D. Williams, D., *Indust. & Engineering Chem.*, 1946, **18:**157-160.

---

[e] 1,1,1,2-tetrachloroethane is a by-product in the production of TeCA (Health Council of the Netherlands. 2006/090SH. The Hague, Netherlands. December 18, 2006, 39).

5   B. Hennig, B., U.S. Patent No. 2,160,574. 1939. 1-2.
6   R. G. Heitz and W. E. Brown, U.S. Patent No. 2,442,324. 1948. 1.
7   F. Richter, *Beilsteins Handbuch der Organischen Chemie*. Vol. 1. Springer-Verlag, Berlin, Germany, 1958, 895.
8   National Cancer Institute, DHEW Publication No. 76-802. U.S. Department of Health. Education and Welfare. National Institute of Health, Bethesda, MD. Appendix A: Chemistry, 1976, 43-68.
9   E. M. Waters, Gerstner, H. B. and J. E. Huff, *J. Toxicology & Environ. Health.* 1977, **2**(3):671- 707.
10  D. W. Hardie, in *Kirk-Othmer Encyclopedia of Chemical Technology*, 2$^{nd}$ Edition. Volume 5, Interscience Publishers, New York, NY. 1964, 159-164.
11  M. Luotamo and V. Viihimaki, Nordic Council of Ministers, DECOS and NEG Basis for an Occupational Standard, 1996, 28:3.
12  F. Zollinger, *Atch Gewerbepathol Fewerbehyg*, 1931, **2**:298-325.
13  California Environmental Protection Agency, Public Health Goal for 1,1,2,2-Tetrachloroethane in Drinking Water, 2003, 3.
14  U.S. Environmental Protection Agency, Regulatory Determinations Support Document for Contaminant Candidate List (CCL2). EPA Report 815-R-08-012, 2008, 11-3.
15  U.S. Department of Health and Human Services, NTP Technical Report on the Toxicity Studies of 1,1,2,2-Tetrachloroethane Administered in Microcapsules in Feed to F344/N Rats and B6C3F$_1$ Mice. National Toxicology Program. Toxicity Report Series No. 49, 2004, 7.
16  J. S. House, Enhanced bioremediation of 1,1,2,2-tetrachloroethane in wetland soils. M.S. Thesis, Louisiana State University, Baton Rouge, LA, 2002, 2-15.
17  World Health Organization, 1,1,2,2-Tetrachloroethane. Concise International Chemical Assessment Document 3. Geneva, Switzerland, 1998, 5.
18  Agency for Toxic Substances and Disease Registry, Toxicological profile for 1,1,2,2-tetrachloroethane. U.S. Department of Health and Human Services, Atlanta, GA, 2008, 111-114.
19  J. F. Pankow and J. A. Cherry, in *Dense Chlorinated Solvents and Other DNAPLs in Groundwater: History, Behavior and Remediation*, Waterloo Press, Portland, OR, 1996, 509.
20  National Institute for Occupational Safety and Health. Occupational Exposure to 1,1,2,2 Tetrachloroethane. NIOSH Publication No. 77-121. Appendix IV, 1976, 143.
21  UNEP, 1,1,2,2-Tetrachloroethane. SIDS Initial Assessment Report. Boston, MA, 2002, 5.
22  W. A. Arnold, W. P. Ball and A. L. Roberts, *J. Contam. Hydrol*, 1999, **40**:183-200.
23  I. Mochida, H. Noguchi, H. Fujitsu, T. Seiyama and K. Takeshita, *Can. J. Chem.*, 1997, **55**:2420-2425.
24  C. A. Schanke and L. P. Wackett, *Environ. Sci. & Technol.*, 1992, **26**:830-833.
25  M. H. A. van Eekert, A. J. M. Stams, J. A. Field and G. Schraa, *Appl. Microbiol. Biotechnol*, 1999, **51**:46-56.
26  T. M. Vogel, C. S. Criddle and P. L. McCarty, *Environ. Sci. & Technol.*, 1987, **21**:722–736.
27  M. M. Lorah, L. D. Olsen, B. L. Smith, M. A. Johnson and W. B. Fleck, Natural Attenuation of Chlorinated Volatile Organic Compounds in a Freshwater Tidal Wetland,

Aberdeen Proving Ground, Maryland. U.S. Geological Survey Water-Resources Investigations Report 97– 4171, 1997, 95.
28 C. Chen, J. A. Puhakka and J. F. Ferguson, *Environ. Sci. & Technol.*, 1996, **30**:542–547.
29 M. Lorah and L. Olsen, *Environ. Sci. & Technol.*, 1999, **33**:227-234.
30 J. Dolfing, *Environ. Sci. & Technol.*, 1999, **33**:2680.
31 D. Hunkeler, R. Aravena, K. Berry-Spark and E. Cox, *Environ. Sci. & Technol.*, 2005, **39**:5975-5981.
32 U.S. Environmental Protection Agency, 1,1,2,2-Tetrachloroethane. Chpt. 11. Regulatory Determinations Support for Selected Contaminants from the Second Drinking Water Contaminant Candidate List (CCL2). USEPA Rpt. 815-R-08-012. Washington, D.C., 2008, 11-3.
33 Canadian Environmental Protection Agency, 1,1,2,2-Tetrachloroethane. Priority Substances, List Assessment Report. Environment Canada, 1993, 4.
34 Canadian Council of Ministers of the Environment, Chlorinated Ethanes. Canadian Water Quality Guidelines for the Protection of Aquatic Life, 1999, 1.
35 U.S. Environmental Protection Agency, EPA/635/R-09/001F, 2010, 4.
36 C. B. Shepherd in *Chlorine Its Manufacture, Properties and Uses,* ed. J. S. Sconce, American Chemical Society Monograph Series. Reinhold Publishing Corporation, New York, NY, 1967, 375-428.
37 O. W. Cass, U.S. Patent 2,280,794, 1942, 1-5.
38 G.W. Warren, U.S. Patent 2,577,388, 1951, 1-10.
39 R. Feathers and R. Rogerson, U.S. Patent 2,914,575, 1959. 1-4.
40 A.C. Ellsworth and R. M. Vancamp, U.S. Patent 2,951,103, 1960, 1-6.
41 U.S. Environmental Protection Agency, Organic Chemical Manufacturing. Vol. 8: Selected Processes. EPA-450/3-80-028c. Office of Air Quality Planning and Standards, Research Triangle Park, NC. 1980, II-1.
42 M. M. Lorah, M. A. Voytek, J. D. Kirshtein and E. J. Jones, Water Resources Investigations Report 02-4157. United States Geological Survey, Aberdeen Proving Ground, Maryland, 2003, 61.
43 N. P. Page and J. L. Arthur, NIOSH Publication No. 78-130. National Institute for Rockville, Occupational Safety and Health (NIOSH). U.S. Department of Health, Education and Welfare. MD, 1978, 10.
44 J. C. Ochsner, T. R. Blackwood and W. C. Micheletti, EPA-600/2-79-210m. Industrial Environmental Research Laboratory. Cincinnati, OH, 1979, 26.
45 Stanford Research Institute, Summary Plant Observation Report and Evaluation for 1,1,2,2-Tetrachloroethane. NIOSHTIC 00068143. Submitted Under NIOSH Contract No. 099-74-0031, 1976, 1.
46 U.S. Environmental Protection Agency, EPA-600/2-2-79-210m, 1979, 7.
47 U.S. Environmental Protection Agency, EPA-560/11-79-009, 1979, 46.
48 U.S. Environmental Protection Agency, EPA-440/4-85-014, 1981. 3-1 to 3-2.
49 U.S. Department of Health, Education and Welfare, NIOSH Publication No. 77-121, 1976,19.
50 W. U. Seiler, *Chem. Eng. News,* Commodity Survey, May 30, 1960, 127.
51 F. A. Loweinheim and M. K. Moran in *Industrial Chemicals,* 4th Ed. Wiley Interscience, Inc. New York, NY, 1975, 604-611.

52. S. A. Cogswell, in *Chemical Economics Handbook*, Stanford Research Institute, Menlo Park, CA, 1978, 632.3000A-F and 632.3001A-632.3002A.
53. A. Goldfarb, W. Duff, E. Herrick and M. McLaughlin, MITRE Corp. MTR-80W27. McLean, VA, 1980, 56-77.
54. ERG Memorandum, April 14, 2006 from Lori Weiss of ERG to Samantha Lewis of the U.S. EPA, 11.
55. R. D. Morrison and B. L. Murphy in *Chlorinated Solvents A Forensic Evaluation*. Royal Society of Chemistry, Cambridge, UK, 2013, 187-194.
56. J. E. Johnson, H. R. Baker, D. E. Field, F. S. Thomas and M. E. Umstead, NRL Report 6197. U.S. Naval Research Laboratory, Washington, DC, 1964, 21.
57. R. A. Saunders, Naval Research Laboratory Report 6206. U.S. Naval Research Laboratory, Washington, DC, 1965, 1-7.
58. R. Stewart, H. Gay, A. Schaffer, D. Erley and V. Rowe, *Arch. of Environ. Health*, 1969, **19**:467-474.
59. D. Henschler, W. Romen, H. M. Elasser, D. Reichert, E. Eder and Z. Radwan, *Int Arch Occup Environ* 1980, **43**:237-248.
60. C. Malton, C, Cotti and V. Patella, *Act Oncologica*, 1986, **7**:101-117.
61. R. Holt, Physical properties of contaminated TCE and 1,1,1-trichloroethane. Technical Communications. Allied Signal Aerospace Company, Kansas City, MO, 1990, 87.
62. T. K. Mohr, J. A. Stickney and W. H. DiGuiseppi in *1,4-Dioxane and Other Solvent Stabilizers*. CRC Press, Baton Raton, FL, 2010, 436.
63. U.S. Environmental Protection Agency, EPA-440/4-85-009, 1982, A-18.
64. G. I. Gruber, Organic Chemicals, Pesticides and Explosives Industries, USEPA. Contract No. 68-01-2919, 1976, 5-34.
65. R. E. Doherty, *J. Environ. Forensics*, 2000, **1**:84.
66. J. Borror, and E. E. Rowe, U.S Patent No. 4,293,433, 1981, 1.
67. H. Tsuruta, K. Iwasaki and K. Fukuda, Letter to the Editor. *Industrial Health*, 1983, **21**:295.
68. World Health Organization, Environmental Health Criteria 50, International Programme on Chemical Safety. Geneva, Switzerland, 1985, 2.2.2.
69. G. M. Carlton, Utilizing air stripping technology for pretreatment of solvent waste in *Solvent Waste Reduction Alternatives Symposia Conference Proceedings*, 1986, 17.

# A FORENSIC METHODOLOGY FOR INVESTIGATING PCE RELEASES FROM A DRY CLEANING FACILITY

Robert D. Morrison

Independent Consultant. 56-2773 Lahuiki Place, Hawi, HI. 96719, USA

## 1. INTRODUCTION

In 2001 it was estimated that approximately 75% of all dry cleaning facilities in the United States (U.S.) were contaminated, primarily with perchloroethylene (PCE).[1] In a 2002 survey of 26 dry cleaning sites in California, PCE was detected at 85% of the sites.[2] In 2008, the U.S. EPA estimated that 15,750 active dry cleaners required remediation while a Florida study of 150 contaminated dry cleaner sites revealed that contaminated groundwater had migrated off-site at approximately 57% of the facilities. Regardless of the actual number of PCE contaminated dry cleaning facilities in the U.S., their investigation and remediation will require decades and consume billions of dollars.[a]

A challenge in the forensic investigation of dry cleaning facilities is the reconstruction of a PCE release after the facility has undergone remediation and/or no opportunity exists for sample collection. A more generic challenge is the absence of standardized forensic investigative procedures. While standards for conducting traditional environmental investigations are available,[3,4,5] an accepted *forensic* investigative methodology specific to dry cleaners is not. [b,c]

---

[a] The Clean Air Act of 1990 identified PCE as a hazardous air pollutant while The World Health Organization and the U.S. EPA have classified PCE as *probably carcinogenic to humans* and *likely to be carcinogenic in humans by all routes of exposure*, respectively.

[b] A distinction is made between a *traditional environmental investigation* whose objective is to identify the vertical and horizontal extent of the contamination (USEPA, Guidance for Conducting Remedial Investigations and Feasibility Studies Under CERCLA, EPA/540/G-89/004, 1988, 1-3) and a *forensic investigation* that identifies the source and age of a contaminant release, which is often used in cost allocation negotiations (R. D. Morrison and B. L. Murphy in *Environmental Forensics Contaminant Specific Guide*, Academic Press, Oxford, 2006, Foreward*)*.

[c] An exception is Chapter 10 titled Forensic Investigations of Dry Cleaners *in Chlorinated Solvents A Forensic Evaluation* published by the Royal Society of Chemistry in 2013 by R. D. Morrison and B. L. Murphy, 325-350.

## 2. FORENSIC METHODOLOGY

The forensic investigation of PCE contamination at a dry cleaning facility includes many of the components of a traditional environmental investigation. [6,7,8] While every investigation is unique, a proposed forensic investigative methodology includes the following components:
- Collection of current and historical site specific operational procedures and equipment information;
- Identification of non-dry cleaner PCE sources;
- Identification and quantification of PCE background concentrations;
- Identification of site-specific PCE sources;
- Selection of sampling locations, density and media;
- Selection of an appropriate forensic analytical program;
- Sample and data reliability analysis, and
- Exploratory data analysis (EDA), if appropriate.

### 2.1 Collection of Site Specific Operational Procedures and Equipment Information

Ideally, the collection of operational and equipment information includes a chemical use and equipment chronology, [d] regulatory and operational documentation, the historical location(s) and generation of dry cleaning equipment, [e] drain and sewer information, solvent usage practices and handling procedures, a chronology of facility renovations, improvements and building expansion details, solvent mileage records,[f] the current and historical location of dumpsters and/or solvent storage areas, groundwater properties and a dendroecological[g] assessment.[9]

*2.1.1 Chemical Usage and Equipment Chronology*
The historical use of dry cleaning solvents and/or type of dry cleaning equipment (e.g., coin operated facilities used PCE > 97% of the time), can provide insight regarding solvent composition (Figure 1).[10,11]

---

[d] A 2006 California EPA report found that the majority of dry cleaning facilities operate a single dry cleaning machine; when considering the number of facilities with more than one machine, the recommended ratio is 1.091 machine per facility (California EPA, California Dry Cleaning Industry Technical Assessment Report, 2006, IV-3).

[e] This category also includes the historical locations of distillation units, solvent and waste solvent storage locations, spotting boards, vacuum unit(s), boiler(s), air compressor(s), subsurface piping, utility locations, AST/UST locations, lint storage/disposal areas, machine floor bolts and staining. Electrical breaker switch panels can often list specific dry cleaning equipment next to each breaker switch. Sanborn maps can also include the location of dry cleaning machines and/or solvent storage areas.

[f] Solvent mileage is often described as the pounds of PCE used to clean 1,000 pounds of clothing or the pounds of clothes cleaned per gallon of PCE. The term *solvent mileage* (also solvent consumption) replaced the 1930s to 1940s equivalent term *pounds per gallon.*

[g] *Dendroecology* is the temporal study of ecological and environmental changes depicted in tree rings (R. D. Morrison and G. O'Sullivan, Forensic Applications of Dendroecology in *Introduction to Environmental Forensics 3rd Edition*, eds. B. L. Murphy and R. D. Morrison, Academic Press, Oxford, UK, 2015, 532).

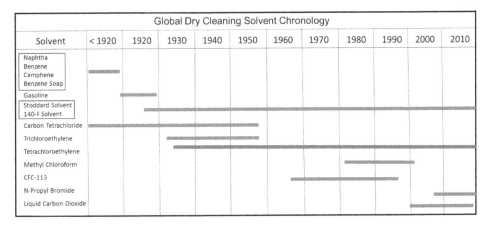

**Figure 1**[h] *Chronological use of dry cleaning solvents*

If the dry cleaning equipment generation, solvent and weight of clothing cleaned is known, the PCE volume used can be estimated. For example, the PCE (gallons) used per 1000 pounds (lbs) of clothing cleaned for five generations of dry cleaning machines is summarized in Table 1.[12] Another approach is to estimate the total PCE lost by the generation of the dry cleaning equipment and/or to assign an assumed leakage factor based on the assumed solvent mileage.

*2.1.2 Regulatory and Operational Documentation*
Regulatory documentation includes equipment and building permits and plans, business permits, training certificates, air quality emission permits, equipment permits and disposal records. Air quality emission permits can include equipment manufacturer and/or model designations. Documented violations of municipal, state and/or federal laws and site inspection records often provide useful operational information.

Operational information includes business hours, number of machines and employees, business status, annual receipts from the operation, percent annual receipts from dry cleaning services only, types of services provided and years operated. Operational information is most commonly used for identifying potential contaminant source areas as well as possible foundation information for developing a cost allocation between multiple parties using a throughput analysis.

---

[h] Figure 1 is a general portrayal of solvent usage; start and end dates vary between countries and often in the literature. Commercial products can also include multiple solvents. For example, in Japan in 1992, a common dry cleaning solvent consisted of 70% petroleum hydrocarbons, 5% chlorofluorocarbons, 5% methyl chloroform and 20% PCE. (J. Kubota, Dry cleaning in Japan: current conditions and regulations, in *Proceedings, International Roundtable on Pollution Prevention and Control in the Dry Cleaning Industry*, EPA/774 R-92/002, 1992, 40-43).

**Table 1** *Solvent mileage for five generations of dry cleaning equipment.*

| Equipment Type | Generation & Date Introduced | Clothing (lbs) Cleaned per Gallon of PCE | PCE (lbs) used per 1000 lbs of Clothing Cleaned |
|---|---|---|---|
| Transfer [a] | 1st - 1930 | 85-128 | 12.2-16.3 |
| Vented Dry to Dry [b] | 2nd - late 1960s | 95-176 | 7.8-14.6 |
| Closed Loop/Dry to Dry (non-vented) [c] | 3rd - late 1960s | 254 | 5.5 |
| Closed Loop Dry to Dry (non-vented) [d] | 4th - early 1990s | 300 | 4.62 |
| Closed Loop Dry to Dry (non-vented) [e] | 5th - late 1990s | 300 | 4.62 |

[a] Also known as a *cold machine* [b] A refrigerated condensation unit where all the exhaust vapors are vented to the atmosphere when the dry cleaning machine door is opened. With a vented system, control of about 85 percent of the solvent vapors is achieved compared to an uncontrolled machine. [c] A closed-loop (dry-to-dry) dry cleaning machine equipped with a refrigerated condenser. [d] A non-vented, closed loop process machine (dry-to-dry) includes an additional internal vapor recovery device. Refrigerated condensers and carbon adsorbers are included in these machines. [e] A non-vented, closed loop dry to dry machine with carbon adsorption and refrigerated condensers is used to reduce residual solvent in the machine cylinder at the end of the dry cycle to concentrations generally less than 300 to 100 parts per million. Fifth generation machines have inductive fans and lockout devices that do not allow the machine door to be opened until solvent vapor levels are reduced to low levels.

*2.1.3 Historical Location and Generation of Dry Cleaning Equipment*
The historical location and generation of dry cleaning machines can provide a basis for assigning a particular geographic area of contamination with a process, solvent, tenant and/or time frame. For example, the generation of the dry cleaning equipment can provide a basis for developing allocation models while the manufacturer model and year of construction can be used to estimate the percentage of PCE in the sludge generated by specific equipment (Table 2).[13]

**Table 2** *Percentage of PCE in sludge from different models of primary and secondary machines.*[a]

| Model and Year | Number of Tests | Sludge Density (g/ml) | PCE (%) in Sludge |
|---|---|---|---|
| Victory 1986 [s] | 1 | 1.15 | 44 |
| Victory 1996 [s] | 2 | 1.17-1.20 | 38-45 |
| Midwest 1988 [p] | 3 | 1.2-1.37 | 41-69 |
| Columbia 1993 [p] | 3 | 1.17-1.30 | 39-65 |
| Columbia 1997 [s] | 3 | 1.19-1.20 | 41-51 |
| Columbia 2000 [p] | 3 | 1.09-1.18 | 31-40 |
| Bowe Permac 1991 [p] | 3 | 1.09-1.08 | 11-21 |
| Bowe Permac 1994 [p] | 3 | 1.08-1.20 | 17-44 |
| Bowe Permac 1999 [p] | 2 | 1.02-1.16 | 14-19 |

[a] Primary machines (p) are closed-loop machines with primary control while secondary machines (s) are closed-loop machines with both primary and secondary control.

For dry cleaning equipment, the manufacturer and model number, the drum size, the location(s) of the washer/dryer, distillation still(s) and/or sludge cooker(s) are important. With this information, a site reconnaissance visit prior to drafting a sampling plan is optimized, especially for confirming and identifying facility features, including bolt holes, concrete patches, sumps, ventilation ducts, electrical panels and floor drains, all of which may be germane to identifying potential sources of a PCE release. Geographic Information System (GIS) mapping of flooring features is useful (i.e., the presence of coffee-colored floor stains, can indicate a boil over from distillation equipment).

Absent direct evidence regarding the dry cleaning equipment, categorization of the operation as coin operated, commercial and/or industrial may provide this and additional useful information. For coin operated dry cleaners, Standard, Inc., a small company in Dallas, TX, first introduced a commercial coin operated dry cleaning machine in 1959, followed by Norge,[i] a division of Borg-Warner, in 1960 and Whirlpool in the same year.[14] In some regions of the U.S., PCE storage in underground storage tanks (USTs) were associated coin operated dry cleaning machines installed in the 1960s.[5] In 1976, it was estimated that coin operated dry cleaning facilities usually consisted of two or three small dry cleaning machines (8-12 lb of clothing capacity) with an annual throughput of 19,811 lbs of clothes. In 1992, virtually all coin-operated machines used PCE.[15] Coin operated dry cleaning machines were banned in California on December 21, 1994.[2] As shown in Table 3, facility categorization can also provide information regarding the most likely solvent used at the facility.[16]

Commercial dry cleaning facilities cleaners include franchise shops offering non self-service cleaning, such as *One Hour Martinizing* and specialty services for leather and other fine goods cleaning. In 1978, there were an estimated 15,060 commercial cleaners in the U.S. with 73% using PCE while the remainder used petroleum solvents (24%) or CFC-113[j] (3%).[16]

**Table 3** *Number of dry cleaning operations and solvents used in the United States in 1986*

| Solvent | Coin Operated | Commercial | Industrial | Total |
|---|---|---|---|---|
| PCE [a] | 4,300 | 14,348 | 251 | 18,889 |
| CFC-113 | -- | 489 | --- | 489 |
| Methyl Chloroform [b] | -- | 50 | -- | 50 |
| Petroleum Solvents | -- | 1,418 | 931 | 2,349 |
| Total | 4,300 | 16,305 | 1,182 | 21,787 |

[a] According to a 1992 reference, coin-operated facilities only used PCE (E. Linak, A. Leder and Y. Yoshida, in $C_2$ *Chlorinated Solvents in Chemical Economics Handbook,* Menlo Park, CA, SRI International, 1992, 632.3000-632.3001) while a 2010 reference cites the use of PCE and CFC-113 (B. Linn et al., Conducting Contamination Assessment Work at Dry Cleaning Sites, State Coalition for Remediation of Drycleaners, 2010, 5). [b] Methyl chloroform data is from 1988.

---

[i] Dow Chemical formulated a special PCE based solvent, Norge-Clor, made to Norge specifications containing certain undisclosed additives (C&EN, October 17, 1960, 38).
[j] CFC-113 = 1,1,2-trichloro-1,2,2-trifluoroethane; $Cl_2FC-CClF_2$.

Industrial cleaners often provide rental services items such as uniforms, mops[k] and mats to industries and institutions. In 1979, the typical industrial dry cleaning plant had one dry cleaning system with an annual throughput of 530,000 to 1,500,000 lbs of clothes with about 50% of these facilities using PCE.[16]

*2.1.4 Drain and Sewer Information*

Prior to 1986, dry cleaners could legally discharge condensate waters into sanitary sewers.[17] The location and composition[l] of sewer pipes and drains, their construction details,[m] direction of flow, pipe length, diameter and thickness, coupling interval, backfill material composition, locations of 90 and 45° joints, cleanout locations and manholes, line maintenance records[n] and lateral and trunk line connections is useful information to acquire. The acquisition of any rodding, balling or jetting records is also important as sewer maintenance activities potentially affect the integrity of the colmation layer [o] thereby potentially increasing the exfiltration rate. If sewer camera records are available, sewer defects (e.g., crack, fracture, break, etc.), extent (e.g., width and length of a crack) and the orientation may be discernable that can be used for identifying subsequent sampling locations to demonstrate the release of PCE.

The premise for acquiring and mapping drain and sewer information is that PCE leakage as a dense non-aqueous phase liquid (DNAPL), vapor and/or dissolved phase from the sewer is likely. Prior to the 1980s, it was permissible for dry cleaning plants to discharge condensate wastewater laden with up to 150 parts per million (ppm) of PCE to sanitary sewers.[p] In the 1992 study by the California Central Water Quality Control Board (CCWQCB), the discharge point for the majority of the dry cleaners was the sewer. This observation was consistent with site inspections and dry cleaning equipment manuals.[18]

*2.1.5 Solvent Usage and Handling Procedures*

The historical volume of PCE and/or other solvents (including spotting agents and detergents) purchased and consumed and information regarding the supplier/vendor is frequently useful information. The method of solvent delivery, transfer and storage is important. Barrel storage areas, underground storage and above ground storage tank locations, delivery truck or railroad

---

[k] Mop water from cleaning the floor at a dry cleaning facility can be saturated with solvent ( B. Linn, State Coalition for Remediation of Drycleaning Sites, CLU-IN Internet Seminar, June 8, 2011, 35). The handling and disposal history of mop water may therefore be of interest in identifying potential PCE source areas.

[l] Materials of construction can include iron, porcelain drains, vitrified clay pipe, polyvinyl chloride (PVC) and acrylonitrile butadiene styrene (ABS).

[m] Construction features include pipe depth and connecting design, such as bell and spigot, buttressed, compression joints or glued joints.

[n] Maintenance records can include historical visual inspection, smoke test results, hydrostatic pressure and isolation testing, lines flushing and repair/replacement history and pressure tests.

[o] Relative to sewers, a *colmation layer* is the transition zone between the sewage in the pipe and the backfill material surrounding the pipe (R. D. Morrison in *Proceedings of the 2013 INEF Conference*, ed. R. D. Morrison and G. O'Sullivan, St. John's College, Royal Society of Chemistry Publishing, Cambridge, 2014, 1-25).

[p] Warm water and/or detergents in separator water with PCE can increase its solubility limit (~150-200 mg/l). PCE and water emulsions may also be formed due to vibration, the presence of lint, organic material, spotting chemicals, moisture, heat, detergents or soaps in PCE/water mixtures, rapid boiling of PCE in the still, uncondensed PCE vapors bubbling through the solvent separator, and intimate mixing of PCE and water in a vapor adsorber.

unloading areas and procedures for solvent transfer can define a geographic area for forensic sampling. In 1992, it was estimated that in Holland, 50 percent of soil and ground-water contamination problems from dry cleaning operations were traced to improper or sloppy transfer operations.[19] In a study of leaks, spills and discharges at Florida dry cleaner sites, approximately 15.3% were associated with solvent transfer and storage activities. The majority of these releases were related to solvent delivery and transfer (over 85% of the reported incidents), most commonly by tanker truck, or filling the dry cleaning machine with solvent.[20]

*2.1.6 Chronology of Facility Renovations, Improvements and Building Expansions*
Facility chronology information includes, building and flooring additions and/or replacements, electrical wiring and drainage modifications, concrete patching, re-painting or renovations due to fire and/or flooding, the presence of expansion joints/cracks near the service door(s) and the location of current and former solvent storage areas. This information if often key for identifying optimal locations for sample collection.

*2.1.7 Solvent Mileage Records*
When a facility commences operation and the solvent tanks are filled, the reference volume is recorded on the solvent mileage log, along with subsequent additions. The solvent mileage log can provide a basis for allocating potential liability between tenants/operators based on solvent usage. A solvent mileage log can also include the weight of each load that is dry cleaned, unusual loads, solvent additions and unusual occurrences such as spills, leaks and/or maintenance repairs.

Mohr (2002) introduced a dry cleaner age-duration threat ranking based on an assumed solvent mileage for each of machine derived from the lifetime solvent mileage of the dry cleaning operation.[12] An *assumed leakage factor* was included in the ranking for 1947-1965 (3%), 1965 (2%), 1976-1985 (1%), 1986-1992 (0.5%) and 1992-2000 (0.1%). While the leakage factor is not intended to be representative of actual releases, it did provide a means to develop a ranking based on solvent mileage.

*2.1.8 Historical Dumpster Locations*
The historical location of dumpsters identified through aerial photographs, waste contractor records or witness interviews represent potential PCE release sources, due to the historical disposal of PCE enriched filters, sludge, lint and soil. The historical location of dumpsters and/or type of used filters plotted through time can be used to associate a tenant with a particular dumpster and PCE source area.

*2.1.9 Groundwater Properties*
Properties important for characterizing the groundwater at a dry cleaning facility includes the historical depth, direction of flow, influence of pumping wells on groundwater flow and temporal fluctuations in the water table elevation especially if it contacts or submerges a portion of the sewer system, and the presence or absence of a DNAPL. A survey of dry cleaning case profiles in the U.S., indicated that 38 to 48% had presumptive evidence of the

presence of DNAPLs in the subsurface.[q] Direct and indirect techniques for locating and identifying DNAPLs in the subsurface are available in the literature.[21]

*2.1.10 Dendroecology Assessment*

A forensic investigation incorporating dendroecology may confirm the presence of PCE in shallow groundwater prior to the collection of samples.[22,23,24,25,26,27] In order to ascertain whether this is a viable technique, trees along the down-gradient axis and periphery of the PCE plume during the time frame of interest is mapped, often with aerial photography, to establish their presence during the period that the dry cleaner operated. Identification of the tree species is important[r] as the root systems require groundwater contact, which is usually $\leq$ 30 ft.[28,29] Dendroecology can also identify the leading edge of a contaminant plume and/or discriminate between PCE sources using compound specific isotope analysis (CSIA)[s] of tree core samples.

**2.2 Identification of Non-Dry Cleaning Related PCE Sources**

Identification of PCE sources other than the dry cleaning facility of interest (or others in the immediate vicinity) is important, especially for allocations considerations. The results of this evaluation can provide a template for performing directed forensic sampling and analysis to discriminate between non-dry cleaner related PCE sources.

**2.3 Identification and Quantification of Background PCE Concentrations**

Standardized procedures are available for the collection, quantification and comparison of background samples with facility samples.[30] For PCE, the task is somewhat easier as PCE is anthropogenic in origin with no naturally occurring sources.[31] Background PCE sources, other than for dry cleaning, can include metal cleaning and degreasing operations, particularly for cleaning aluminum prior to the development of stabilized methyl chloroform formulations, and for the removal of wax and resin residues. Other uses and products include automotive brake cleaning, an ingredient in pesticide formulations prior to about 1985/1988 in the U.S., rubber dissolution, film cleaning,[32] as an ingredient in adhesive coating,[33] paint removal, sulfur recovery, a grain fumigant, soot removal, and catalyst regeneration. The primary use of PCE after 1996 was for the production of fluorinated compounds, such as CFC-113 and HFC-134a.[t] By 2004, aerosol products (for cleaning tires, brakes, engines, carburetors, wire) constituted 12% of the total use of PCE. PCE was also used as a dielectric fluid in

---

[q] In a forensic investigation, the purpose of identifying the location and/or most probable location of a DNAPL is for the purpose of collecting a sample with the highest probability of containing age diagnostic impurities, detergents, and/or stabilizers as contrasted with its collection for contaminant characterization and remediation purposes (See footnote b).
[r] In addition to differences in root depths between tree species, trees sorb and accumulate contaminants differently. In some tree species, bacterial infestation in decaying wood tissue may provide a dechlorination mechanism within the tree, potentially resulting in a false negative interpretation.
[s] Compound specific isotope analysis (CSIA) is an analytical method that measures the ratios of naturally occurring stable isotopes in environmental samples.
[t] 1,1,1,2-tetrafluoroethane

transformers, capacitors and circuit breakers.[u] Businesses near the dry cleaner(s) investigated can include auto repair shops, junk yards, residences using septic cleaning products containing PCE, precision degreasing operations, semi-conductor and/or aerospace company manufacturers with degreasing activities, solvent storage and distribution companies and textile plants. These businesses are especially important if they share a similar sewer or storm water drainage system with the dry cleaning site, especially if leakage from the piping down gradient of the facility is suspected (e.g., comingled with up gradient and/or site specific releases).

## 2.4 Identification of Site Specific Sources

Relevant studies of site specific sources at dry cleaning facilities include a Florida Dry Cleaning Survey and a California Regional Water Quality Control Board (RWQCB) report. Data collected by the Florida Dry Cleaning Solvent Cleanup Program in the 1990s examined reported spills, leaks and discharges of dry cleaning solvent from 334 dry cleaning facilities and 14 dry cleaning wholesale businesses in Florida.[20] Of the active facilities, 31.8% identified at least one solvent leak, spill or discharge. Of 530 reported incidents, most were solvent discharges although others included still bottoms, filters and/or contact water.[v] Discharges associated with operation of the dry cleaning equipment included (a) dry cleaning equipment failure, (b) routine equipment operation, (c) solvent transfer and storage activities, and (d) releases associated with equipment maintenance.

*2.4.1 Dry cleaning Equipment Failure*
The Florida study identified 39.2% of the solvent releases with dry cleaning equipment failure from leaking seals, gaskets, piping, hoses and valves associated with equipment wear and corrosion, expansion and contraction of metal from temperature changes and equipment vibration. Leaking door gaskets were the most common source of releases. Filter discharges were associated with failed seals or leaks of the cartridge filter housing or the accumulation of pressure in cartridge filters that resulted in filter rupture.

*2.4.2 Routine Equipment Operation*
Solvent releases from equipment operator error (20.9%) included boil over of solvent/distillation residues from distillation units (34%), open machine doors while the equipment is operating (9%), loose cartridge filter housing, water separator overflows and open valves (4%).

*2.4.3 Solvent Transfer and Storage*
Solvent storage and transfer incidents accounted for 15.3% of all releases with over 85% associated with solvent transfer by tanker truck to a storage vessel or into the dry cleaning

---

[u] Pentaphen and *n*-methyl pyrrole are diagnostic markers for PCE used as dielectric fluids which do not appear to have been used in dry cleaning formulations.

[v] The 530 releases were grouped into six categories: dry cleaning equipment failure (208), dry cleaning equipment/machine operation (111), solvent transfer or storage (81), equipment/machine maintenance (73), discharges of dry cleaning wastes (50) and other spills/discharges (7).

machine. PCE spills occurred during transfer from the tank truck into a storage tank and via leaking valves, spillage from containers used to fill the dry cleaning machine and/or overfilling. Until the advent of closed-loop delivery systems and machines equipped with filling ports, solvent was introduced through the machine door or through the button trap lid, located at the rear of the machine.

*2.4.4 Releases Associated with Equipment and Machine Maintenance*
Dry cleaning machine and equipment maintenance releases (13.8%) were associated with changing and cleaning filters, distillation still cleaning, solvent pump servicing, filter tank draining and miscellaneous activities. Filter replacement were a common source of PCE releases.[34,35] A common operational practice was to drain spent filters overnight before changing, although it was not always observed, resulting in solvent releases from filters. Spent filters were collected and often stored outside the service door of a dry cleaner facility. The average number of standard, split, and jumbo[w] cartridges used in a primary dry cleaning machine per year for a facility cleaning 44,000 pounds of clothing in the early 2000's was 15, 13 and 7, respectively.[35] PCE releases also occur from the disposal of still bottoms or cooked powder residues during cleanout of distillation units or muck cookers after distillation. The residual solvent in still bottoms or cooked powder residue depends on the efficiency of the distillation operation and the composition of the original material prior to distillation. In 2001, the PCE content (by weight) in still bottoms and cooked powder and distillation residue were between 42 and 75% and between about 56 and 60%, respectively.[35,36,37,38] Non-volatile residues in still bottoms can also include detergents, waxes, dyestuffs, sizing, oil, manufacturing impurities, stabilizers and grease.

A study of 300 California dry cleaners found that the average dry cleaner generates 11.4 gallons of still residue, which contains up to 50% solvent along with 4.3 waste filters per month.[39] Six of these dry cleaning facilities were surveyed in detail; the annual amount of still bottom sludge and waste from filter wastes produced at each facility is summarized in Table 4.[39] The assumption that PCE in sludge and filter residue always originates from a dry cleaning solvent may be incorrect. For example, in 2005 PCE was detected in the distillation sludge at 7 of 8 dry cleaners that used using hydrocarbon based dry cleaning solvents.[40] Although the PCE source was unknown, it was hypothesized to have originated from spotting chemicals or from in fabric previously dry cleaned with PCE.In 1992 the California RWQCB[41] examined PCE releases at 17 dry cleaners and concluded that PCE discharges from dry cleaners into sewers had resulted in soil and groundwater contamination.[x] Separator water samples[y] were collected from 9 dry cleaners after sitting for 24 hours; PCE was found to have separated from some samples containing 30% pure solvent. PCE separator water values ranged from 56 to 1,119,300 ug/l with an average of 151 mg/l.

---

[w] The PCE in standard and split filters is estimated to be 0.5 gallons and as much as 1 gallon in a jumbo cartridge filter.

[x] In 1991, it was estimated that wastewater discharges from dry cleaners to municipal sewers occurred with more than 50% of dry cleaners in the U.S. (Federal Register, National Emission Standards for Hazardous Air Pollutants for Source Categories: Perchloroethylene Emission from Dry Cleaning Facilities, December 9, 1991, Washington, DC, 64382).

[y] Separator water is the wastewater generated from the physical separation of drycleaning solvent and water in a water separator. Separator water is a contact water and therefore contains solvent.

**Table 4** *Results of site visits to six dry cleaning plants in 1988 in Southern California*

| Throughput (lbs/yr) | Solvent Purchases (gal/yr) | Equipment Description [a] | Filter Waste Generated (lbs) | Still Residue Generated (gal) [c] |
|---|---|---|---|---|
| 112,000 | PCE (720) TCA (1,200) CFC-113 (1,200) | Dry to dry/refrigeration (2) Transfer with CA[b] (1) Reclaiming tumblers (3) | 288 | 840 |
| 175,000 | PCE (700) | Dry to dry with salvation (2) | 88 | 540 |
| 400,000 | PCE (400) | Dry to dry with refrigeration (2) | 64 | 240 |
| 288,000 | PCE (1,200) | Dry to dry with CA (3) Dryer (1) | 299 | 420 |
| 168,000 | PCE (170) | Dry to dry with refrigeration (2) | 48 | 162 |
| 2,040,000 | Stoddard Solvent (11,350) | Washer extractor (2) Recovery extractor tumblers (6) | 48 | 34 tons |

[a] Numbers in parentheses are the number of machines at the facility; [b] CA = carbon adsorption. [c] Annual

The RWQCB report concluded that PCE releases from sewers occurred from breaks and/or cracks and via pipe joints and other connections. PCE containing liquid or vapor also leached directly through the walls of the sewer lines. The sewer was the single discharge point for the majority of facilities.

**2.5 Selection of Sampling Locations, Density and Media**

The sample media, number of samples and location is dictated by the goal of the forensic program and is not constrained by the traditional soil, vapor and groundwater samples normally collected. Building materials, wall scrapings, duct wipe samples, pipes samples, sediment and wipe samples from catch basins, paint chips, flooring and/or concrete coring samples, lint and fabric samples, sewer camera videos, gutter samples and/or flooring may all be appropriate. Traditional sampling media may also be appropriate but for a different purpose; for example, a limited soil gas survey for a unique diagnostic volatile compound may be appropriate due to the lower achievable detection levels in comparison to a comprehensive soil vapor contaminant characterization study. Other forensic investigative opportunities may include sewer pressure tests/backflushing, excavation and testing of soil around sewer joints, expansion joint and drywall sampling as indicators of floor releases, and sediment and wipe samples from trash dumpsters used for filter cartridge disposal. For forensic investigations, duplicate or triplicate samples may not be required if a single sample examined microscopically, for example, is sufficient to ascertain the presence or absence of a feature (e.g., carbon filter material from a filter, lint, etc.) without the need for an elaborate statistically designed sampling plan.

## 2.6 Selection of an Appropriate Forensic Analytical Program

A forensic analytical plan often diverges dramatically from a traditional dry cleaner characterization due to the differences in their goals. While not exhaustive, several techniques not normally considered are presented, including dioxin and furan congener analysis, PCB congener and homologue pattern recognition, nanotextile analysis and stable isotope testing, an illustrative of this concept.

*2.6.1 Dioxin and Furan Congener Analysis*

The term *dioxin* refers to chlorinated dibenzo-*para*-dioxins (CDDs) while *furans* refers to chlorinated dibenzofurans (CDFs).[z] In some cases, a unique surrogate opportunity exists to distinguish between different PCE release scenarios based upon CDD/CDF congener and/or homologue profiling (e.g., distinguishing between a release from a distillation/muck sludge and virgin product) in an environmental sample. This example assumes that a release of new solvent contains small amounts of CDD/CDF's while distillation sludge is enriched with CDD/CDF's. In a 1990 study of dry cleaning solvent distillation residues collected from 12 commercial and industrial operations, considerable amounts of CDD/CDFs) with the dominant congeners consisting of octachlorodibenzo-*p*-dioxin (OCDD) and heptachlorodibenzofuran (HpCDD) were detected. [42]

In 1999, the United Nations Environment Programme concluded that CDD/CDFs were present in sludge from dry-cleaning units in Germany.[43] In Japan, a four year study in 2001 detected dioxins in PCE, petroleum solvents, spot cleaning solvents, distillation sludge and in new and dry cleaned clothing.[44] Clothing dyes were postulated as a likely dioxin source. The availability of congener and homologue data therefore allows the use of dioxin and furan pattern recognition with PCE in various fabrics, as depicted in Figure 2.

The hypothesis that dioxin enrichment is attributed to the presence of dioxin in clothing was tested by measuring the dioxin in wastewater from a dye factory located in the Nagano Prefecture in Japan which detected the following dioxin and furan concentrations: PCDDs (103 pg/l), PCDFs (74 pg/l), and $\sum$ PCDFs + PCDDs (180 pg/l). A petroleum-based solvent used for cleaning four articles was also tested, with similar results, indicating that the dyes in the clothing was the source of the dioxins.

By 2010, studies in Ireland had concluded that the main source of dioxins from the dry cleaning process originated from the clothing, which contained dioxin.[aa] No dioxins were

---

[z] CDD and CDFs are triple ringed structures consisting of two benzene rings connected by either one (furan) or two oxygen (dioxin) molecules. For dioxins, there are 75 different congeners (chlorine patterns around the aromatic rings) with between one and eight chlorine atoms on the aromatic rings (carbon ring structure wherein the electrons are delocalized over the ring). Dioxins contain two of these aromatic rings. The term *congener* is defined as a compound with multiple configurations with a common chemical structure. A *homologue* is a compound belonging to a family of chemicals differing from each other with a repeating unit, such as a methylene group. For dioxins and furans, the term homologue refers to the same number of chlorine atoms, regardless of position; for tetrachlorodibenzo-*p*-dioxins (TCDD), for example, there are 22 possible isomers within the TCDD homologue class.

[aa] This observation is consistent with a study by the University of Bayreuth in Germany which showed that sludge residues from the distillation unit of a dry cleaning facility using PCE contained dioxins. The data suggested that the CDD/CDF's could have originated from pentachlorophenol that is sometimes used as a preservative for cotton textiles and/or from Chloranil dyes. The distillation process was excluded as a source of

generated during the dry cleaning process but rather dioxins from clothing were concentrated in the cleaning solvents. As the solvents were distilled for recovery and reuse, dioxins were further concentrated in the distillation residues.[45]

*2.6.2 PCB Congener and Homologue Pattern Recognition*

In some circumstances, the analysis of PCBs (polychlorinated biphenyls) with chlorinated solvents in an environmental sample can provide useful information. In the 2001 study in Japan, analysis of selected PCBs in solvents, spot removers and filters was performed. Figure 3 provides potential forensic opportunities using the presence of PCE and PCB congeners to associate the PCE with specific products or time period, especially if PCB congener information is available.

The presence of 2,3,4,4,5-PeCB and 3,4,4,5-TECB in used petroleum solvent, charcoal and solvent filter suggests that the PCBs originated from sources other than new petroleum solvent and/or spot remover which did not contain these constituents, such as clothing and/or soap.

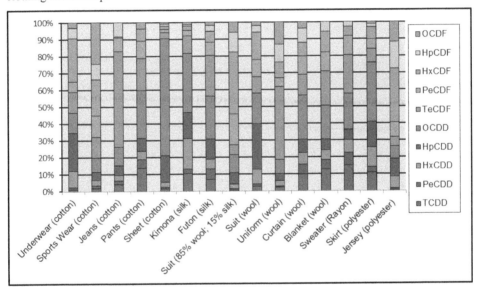

**Figure 2** *Normalized histograms for CDDs/CDFs in 15 new articles of clothing purchased in Japan in 2001 (from R. D. Morrison and B. L. Murphy in Chlorinated Solvents A Forensic Evaluation, Royal Society of Chemistry Publishing, Cambridge, 2013, 345)*

---

dioxin formation since the distillation temperature of about 80°C is below the temperature at which CDD/CDFs are synthesized in thermal processes from precursor organic chemical compounds, such as chlorophenols and chlorobenzenes) in the presence of chlorides (R. Fuchs, J. Towara and O. Hutzinger, *Organohalogen Compounds*, 1991, **3**:441-444).

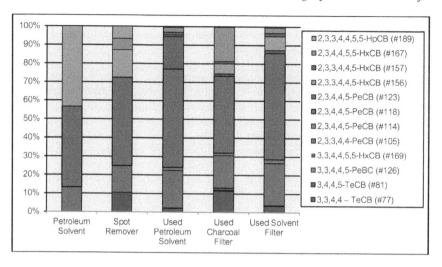

**Figure 3** *Normalized concentration of PCBs congeners from five dry cleaner associated solvents and filters (from R. D. Morrison and B. L. Murphy in Chlorinated Solvents A Forensic Evaluation, Royal Society of Chemistry Publishing, Cambridge, 2013, 348).*

*2.6.3 Forensic Microscopy* [46]
Environmental forensic microscopy [bb] includes the analysis of samples with visible, infrared light and electron microscopy. Light microscopy is usually performed with polarized light microscopy (PLM) but may involve phase contrast (PCM), dark-field or fluorescence microscopy while infrared microscopy is performed using Fourier transform infrared microspectroscopy (FTIR). FTIR is useful when identifying organic molecules such as plastics and polymers. Scanning electron microscopy (SEM) provides the ability to view particles smaller than with light microscopy; when equipped with an x-ray analysis unit, the elemental composition of the particles can be determined. Transmission electron microscopy (TEM) allows viewing of particles smaller than with light microscopy; when equipped with electron diffraction capabilities and an x-ray analysis, the crystal structure and elemental composition of the particles can be examined.

*2.6.4 Nanotextiles*
The presence of nanotextiles with PCE or other diagnostic indicators can provide corroborative information regarding the source and possible age of a release, although its use is presently in its infancy.[47] Of special interest is the proliferation of silver or silver nanoparticles (Ag-NPs) in commercial applications, especially as an antimicrobial, antifungal and antiviral agent in fabrics.[48,49,50] Geranio et al., 2009, for example, examined the amount of

---

[bb] *Environmental forensic microscopy* is the use of microscopy following forensic procedures to characterize particles and materials involved in environmental studies.

silver leached from socks; reported concentrations ranged from 0.3 to 377 ug/g.[cc] One study found that the amount of silver leached from socks was affected by the level of agitation and the fluid and was found primarily in the particulate fraction (>450 nm).[48,4851] The composition and characterization of this coarse fraction includes Ag in fiber fragments, Ag-NPs aggregates or Ag precipitates. Another consideration is that chloride anions from tap water or the detergent may precipitate $Ag^+$ as AgCl, thereby decreasing the amount of dissolved Ag. The use of SEM with an emphasis of identifying silver particles in a sludge, distillation or lint may provide a valuable corroborative tool to identify wastewater discharged from a dry cleaning facility.

*2.6.5 Stable Isotopes*
As with concentration values for chlorinated solvents from groundwater monitoring wells, isotopic values for chlorine, carbon and hydrogen may be used to ascertain whether up gradient sources migrating onto a facility are isotopically distinguishable from on-site values due to differences in the original product and degradation.[dd] If the isotopic signatures of up gradient and down gradient sources of a facility along the groundwater flow path are statistically indistinguishable, the up gradient source is either contaminating the groundwater underlying the facility or releases on the site and up gradient sources shared products which were isotopically similar, such as purchases from the same manufacturer/supplier.

Several compound specific isotope analysis (CSIA) case studies detail its use in identifying and distinguishing between chlorinated solvent sources. Examples include the investigation of groundwater contamination in Orlando, Florida, as originating from a former dry cleaner and uniform rental services businesses, and the use of CSIA to delineate multiple potential sources of PCE and TCE from several dry cleaners, automotive businesses and a former municipal airfield in Leon Valley, Texas.[52,53]

## 2.7 Sample and Data Reliability Analysis

Forensic test results must be reliable. Whether the test results are based on sample specific comparisons, diagnostic ratios or multivariate statistical analyses, the subsequent conclusions are dependent on good quality analytical or forensic data. Analytical data for identical compounds, but from different laboratories or derived by different methods, can introduce uncertainty into the comparisons. The frequency of "non-detects" in the data and how results below the method reporting limit are addressed can potentially bias chemical results, especially at low concentrations. Issues such as co-elution which can lead to the misidentification of analytes or inaccurate quantification, also require examination. Literature regarding data reliability for different testing methods are available and generally follow the

---

[cc] The washing procedure used in the study followed an ISO method for wash-tests using a Washtec-P Roaches washing machine with a speed of 40 ± 20 rpm equipped with steel vessels with a capacity of 550 ml (ISO, *Textile Tests for Colour Fastness spart C06, Colour Fastness to Domestic and Commercial Laundering*, ISO 105-C06; International Organization for Standardization: Geneva, 1997).

[dd] At the time of this writing, error bars for samples with less than about 5 ug/l render the results potentially unreliable.

same steps used to examine environmental data collected from traditional environmental investigations.[54,55]

## 2.8 Exploratory Data Analysis (EDA)

The use of exploratory data analyses provides a significant amount of forensic information for analysis. Mathematical methods can identify patterns (similarities and differences) in groups of data. Multivariate statistical analyses can provide useful information regarding the relative importance of numerous variables affecting an analytical data set. The statistical theory for these techniques is discussed in the literature and includes principal component analysis (PCA), polyvector and discriminant analysis, hierarchical, and other types of cluster analysis.[56]

Of these techniques, principal components analysis has received the most attention in its use for contaminant source identification. Principal components analysis is primarily used to transform sample composition data into smaller and uncorrelated variables called principle components. For data sets with a large number of interrelated variables, principle component analysis is able to provide a basis for analyzing the data structure and reducing the dimensionality of the pattern vectors. The most important variables affecting the data distribution can then be examined.

## 3. CONCLUSIONS

A methodical sequence of activities is needed for conducting an environmental forensic investigation of a dry cleaning establishment to insure that the ability to identify the source and age of a contaminant release is optimized. Traditional contaminant characterization procedures and analysis are frequently of little value in forensic investigations, as the purpose is different. Wherever possible, examine the merit of sampling and testing non-traditional media and whether such sampling contributes to the goal of the forensic investigation.

### References

1. R. Schmidt, R. DeZeeuv, L. Henning and D. Trippler, State Programs to Clean Up Dry Cleaners, State Coalition for Remediation of Drycleaners, Washington, DC, 2001, 2.
2. K. G. Mohr, Study of Potential for Groundwater Contamination from Past Dry Cleaner Operations in Santa Clara County, Santa Clara Water District, Santa Clara, CA, 2002, 99-116.
3. California Environmental Protection Agency, Dry Cleaner Site Discovery Process, 2005, 10.
4. W. Linn, L. Appel, G. Davis, R. DeZeeuw, C. Dukes, P. Eriksen, J. Farrell, D. Fitton, J. Gilbert, J. Haas, L. Henning, R. Jurgens, B. Pyles, R. Schmidt. J. So, A. Spencer, D. Trippler and P. Wilson, Conducting Contamination Assessment Work at Dry Cleaning Sites, State Coalition of Remediation of Drycleaners, 2015, 52.

5   T. Evanson, Site reconnaissance & sampling, in Conducting Contamination Assessment at Drycleaning Sites. CLU-IN Internet Seminar, State Coalition for Remediation of Dry Cleaning Sites. June 8, 2011, 42-65.
6   U.S. EPA, Guidance for Conducting Remedial Investigations and Feasibility Studies Under CERCLA, EPA/540/G-89/004, 1988, 1-3
7   W. J. Shields, Y. Tondeur, L. Benton and M. R. Edwards in *Environmental Forensics Contaminant Specific Guide,* ed. R. D. Morrison and B. L. Murphy, Academic Press, Oxford, 2006, 294-305.
8   R. H. Plumb, Fingerprinting analyses of contaminant data, EPA/600/5-04/054, 2004, 1-4.
9   R. D. Morrison and G. O'Sullivan in *Introduction to Environmental Forensics $3^{rd}$ Edition* ed. B. L. Murphy and R. D. Morrison, Academic Press, 531-546.
10  World Health Organization, IARC Monographs on the Evaluation of Carcinogenic Risks to Humans, Vol. 63, Lyon, France, 1995, 35.
11  R. D. Morrison and B. L. Murphy in *Chlorinated Solvents A Forensic Evaluation,* Royal Society of Chemistry Publishing, Cambridge, 2013, 247-286.
12  K. G. Mohr, Study of Potential for Groundwater Contamination from Past Dry Cleaner Operations in Santa Clara County, Santa Clara Water District, Santa Clara, CA, 2002, 46-47.
13  California Environmental Protection Agency, California Dry Cleaning Industry Technical Assessment Report, 2006, F-1 to F-2
14  Chemical & Engineering News, Norge Launches Coin-Op Dry Cleaner, October 17, 1960, 38.
15  U.S. EPA, Perchloroethylene Dry Cleaners Background Information for Proposed Standards, EPA-450/3-79-029a, 1980, 3-1 to 3-2.
16  Jacobs Engineering Group, Inc., Source Reduction and Recycling of Halogenated Solvents in the Dry Cleaning Industry, Source Reduction Research Partnership, Pasadena, CA, 1992, 4.
17  T. Mohr, *1,4-dioxane and Other Solvent Stabilizers,* CRC Press, Boca Raton, 2012, 27.
18  V. J. Izzo, Dry Cleaners A Major Source of PCE in Ground Water, Central California Region of the Regional Water Quality Control Board, State of California, Rancho Cordova, CA, 1992, 9.
19  J. Cantin, Overview of Exposure Pathways, in *Proceedings of the International Roundtable on Pollution Prevention in the Drycleaning Industry,* EPA/774/R-92/002. Office of Pollution Prevention and Toxics, May 27-28$^{th}$, 1992, Falls Church, VA, Washington, DC, 1992, 5.
20  B. Linn and K. Mixell, Reported Leaks, Spills and Discharges at Florida Drycleaning Sites, http://www.drycleancoalition.org/download/LeaksSpillsandDischarges.pdf., Accessed on 1/22/2015.
21  Interstate Technology and Regulatory Cooperation Work Group, Dense Non-Aqueous Phase Liquids (DNAPLs), Review of Emerging Characterization and Remediation Technologies, Washington, DC, 2000, 3-22.

22. G. Oudijk and J. C. Balouet in *Environmental Forensics Proceedings of the 2009 INEF Annual Conference,* ed. R. D. Morrison and G. O'Sullivan. Royal Society of Chemistry Publishing, Cambridge, 2009, 92.
23. E. M. Sheehan, M. A. Limmer, P. Mayer, U. G. Karlson and J. G. Burken, *Environ. Sci. & Tech.,* 2012, **46**:3319-3325.
24. D. A. Vroblesky, C. T. Nietch and J. T. Morris, *Environ. Sci. & Technol.,* 1998, **33**:510–515.
25. A. Limmer, J. Balouet, F. Karg, D. Vroblesky and J. Burken, *Enviro. Sci. & Technol.,* 2011, **45**:8276-8282.
26. C. J. Balouet, J. Burken, F. Karg, D. Vroblesky, K. Smith, H, Grudd, A. Rindby, F. Beaujard and M. Chalot, *Environ. Sci. & Technol.*, 2012, **46**:9541-9547.
27. M. Larsen, J. Burnken, J, Machackova, U. G. Karlson and S. Trapp, *Environ. Sci. & Technol.,* 2008, **42**:1711-1717.
28. J. G. Schumacher, G. C. Struckhoff and J. G. Burken, Assessment of Subsurface Chlorinated Solvent Contamination Using Tree Cores at the Front Street Site and a Former Dry Cleaning Facility at the Riverfront Superfund Site, New Haven, Missouri, 1999-2003, USGS Scientific Investigations Report 2004-5049, Reston, VA, 2004, 9.
29. S. E. Cox, Preliminary Assessment of Using Tree-Tissue Analysis and Passive Diffusion Samples to Evaluate Trichloroethylene Contamination of Groundwater at Site SS-34N, McCord Air Force Base, Washington, 2001, USGS Water Resources Investigations Report 02-4274, Reston, VA, 2002, 13.
30. Naval Facilities Engineering Command, Handbook for Statistical Analysis of Environmental Background Data, Engineering Field Activity, San Bruno, CA, 1999, 1-137.
31. H. B. Singh. *Geophysical Research Letters,* 1977, **4**:453.
32. National Industrial Chemicals Notification and Assessment Scheme Tetrachloroethylene, Priority Existing Chemical Assessment Rpt. No. 15, Commonwealth of Australia, Sydney, NSW, 2001, 18.
33. U.S. Department of Health and Human Services. Report on Carcinogens, 12th Edition, Public Health Service, National Toxicity Program, 2011, 399.
34. Tennessee Department of Environment and Conservation, Clearing the Air on Clean Air: Strategies for Perc Drycleaners, 1997 Edition, UT Center for Industrial Services, Nashville, TN, 5.
35. California Environmental Protection Agency Air Resources Control Board, California, Dry Cleaning Industry Technical Assessment Report, Sacramento, CA, 2006, IV-23 to IV-24.
36. California Regional Water Quality Control Board, Regional Board Resolution Supporting Dry Cleaner Study Grant Application, Appendix A Chemical and Physical Information, San Francisco Bay Region, San Francisco, CA, 2001, 26
37. Chemische Fabrik Kreussler & Company (no date), Textile Care in Solvent – CLIP – Dry Cleaning Detergents, Wiesbaden, Germany, 9.
38. U.S. EPA, Cleaner Technologies Substitutes Assessment for Professional Fabricare Processes, EPA 744-B-98-001, 1998, 2-7.

39  Jacobs Engineering Group, Inc., Source Reduction and Recycling of Halogenated Solvents in the Dry Cleaning Industry, Source Reduction Research Partnership, Pasadena, CA, 1992, 17-18.
40  M. Morris and K. Wolf, Hydrocarbon Technology Alternatives to Perchloroethylene for Dry Cleaning, Institute for Research and Technical Assistance, Santa Monica, CA, 2005, 56.
41  V. J. Izzo, Dry cleaners – A Major Source of PCE in Ground Water, Central California Region of the Regional Water Quality Control Board, State of California, Rancho Cordova, CA, 1992, 21.
42  R. Fuchs, J. Towara and O. Hutzinger, *Organohalogen Compounds,* 1990, **3**:441-444.
43  United Nations Environment Programme, Dioxin and Furan inventories. National and Regional Emissions of PCDD/PCDF, UNEP Chemicals, Geneva, Switzerland, 1999, 41.
44  T. Fukai, A Study of Dry Cleaning Solvents and Clothing: Source of Dioxin Exposure? Clean Water World Initiative Institute of Sosei Water Cleaning, Nagano, Japan, 2001, 4-18.
45  URS Dames & Moore, Inventory of Dioxin and Furan Emissions to Air, Land and Water in Ireland for 2000 and 2010, Final Report Environmental Protection Agency, Wexford, IR, 2010, 68.
46  J. R. Millette and R. S. Brown in *Introduction to Environmental Forensics 3$^{rd}$ Edition,* ed. B. L. Murphy and R. D. Morrison, Academic Press, Oxford, 2015, 487-508.
47  C. Lorenz, L. Windler, N. von Goetz, R. Lehmann, M. Schuppler, K. Hungerbühler, M. Heuberger and B. Nowack, *Chemosphere,* 2012, **89**:817–824.
48  L. Geranio, M. Heuberger and B. Nowack, *Environ. Sci. & Technol.,* 2009, **43**:8113-8117.
49  J. R. Morones, J. L. Elechiguerra, A. Camacho, K. Holt, J. B. Kouri, J. T. Ramirez, and M. J. Yacaman, *Nanotechnology,* 2005, **16**:2346–2353.
50  D. M. Mitrano, E. Rimmele, A. Wichser, R. Erni, M. Height, and B. Nowack, *ACS Nano,* 2014, **8**:7208-7219.
51  T. M. Benn and P. Westaerhoff, *Environ. Sci. & Technol.,* 2008, **42**:4133–4139.
52  U.S. EPA, Technology News and Trends, 52, Washington, DC, 2010, 4-5.
53  U. S. EPA, Technology News and Trends, 46, Washington, DC, 2010, 1-2
54  U.S. EPA, A Summary of General Assessment Factors for Evaluating the Quality of Scientific and Technical Information, EPA 100/B-03/001, 2003, 1-11.
55  R. H. Plumb, A Fingerprint Analysis of Contaminant data: A Forensic Tool for Evaluation of Environmental Contamination, EPA/600/5-04/054, 2004. 1-27.
56  G.W. Johnson, R. Ehrlich and W. Full in *Introduction to Environmental Forensics,* ed. B. L. Murphy and R. D. Morrison, Academic Press, 2004, 461.

# $\delta^{13}C$ AND $\delta^{37}Cl$ ON GAS-PHASE TCE FOR SOURCE IDENTIFICATION INVESTIGATION - INNOVATIVE SOLVENT-BASED SAMPLING METHOD

Daniel Bouchard[1], Patrick W. McLoughlin[2], Daniel Hunkeler[1] and Robert J. Pirkle[2]

1 Centre for Hydrogeology and Geothermics (CHYN), University of Neuchatel, Rue Emile Argand 11, 2000 Neuchatel, Switzerland
2 Microseeps, a Division of Pace Analytical Energy Services, LLC., 220 William Pitt Way, Pittsburgh, PA15238, USA

## 1. INTRODUCTION

A novel sampling method to collect gas-phase VOC in view to perform compound-specific isotope analysis (CSIA) is presented. By analogy to the sorption tube method, the proposed dissolution tube method uses an organic solvent as a sink to accumulate the VOC. The performance of the sampling device was evaluated at the field scale and was compared to conventional summa canisters. The comparative study was carried out in a former industrial building, where a TCE NAPL source was placed on the floor to create a temporary gas-phase plume. Following the establishment of the TCE gas-phase plume, 3 sampling events were conducted within 24 h. Six (6) summa canisters were placed at two different distances from the source to evaluate uniformity of the isotopic signal across the room and set to different intake rates to evaluate possible fractionation related to the sampling procedure. The solvent-based dissolution method was tested in parallel, and the sampling was conducted at a different positioning from the source compared to the canisters. The $\delta^{13}C$ values for TCE sampled using both sampling devices were found to be similar to the isotopic composition of the NAPL source, independently of the distance from the source. In addition, $\delta^{37}Cl$ value for TCE sampled with the dissolution tubes was also similar to the isotopic composition of the NAPL source. The reliability and the practicability of the solvent-based dissolution method hence offer a promising alternative sampling method to perform 2D-CSIA forensics assessment for gas-phase VOCs.

Compound-specific isotope analysis (CSIA) is an assessment tool increasingly used to perform source differentiation assessment on contaminated groundwater sites.[1-5] Following decades of research for tool application in groundwater studies, the interest of applying CSIA in gas-phase contaminant forensics investigations has recently increased. For instance, CSIA was used to discriminate the origin of volatile organic compounds (VOCs) in investigations related to atmospheric emissions [6-10] or to indoor air studies and vapour intrusion assessments.[11, 12] Due to the likely different $\delta^{13}C$ and $\delta^{37}Cl$ composition of chlorinated compounds produced from different manufacturers [13] or contained in diverse consumer products,[11] CSIA turns out to be an appropriate tool to perform source apportion assessments. Nevertheless, the number of investigations remains limited as the application still represents a significant challenge. The major barrier is the limited mass of VOC that can be collected during sampling due to very low VOC concentration generally observed in indoor or outdoor air.[14] The use of summa canisters or sorption tubes to sample gas-phase VOCs by some of the studies mentioned above demonstrated the possibility to perform reproducible isotopic measurements. Where gas-phase VOC concentrations are low, the use of sorption tubes allows processing a larger air volume in order to accumulate a sufficient VOC mass required for the analysis,[15] and consequently became the preferred sampling method. In a recent laboratory investigation, Bouchard and Hunkeler [16] made use of an organic solvent to accumulate gas phase VOCs in view to perform isotopic measurements. In the latter study, air volumes ranging from 0.25 L to 6 L and containing TCE or benzene were bubbled through methanol or ethanol at different flow rates (from 50 to 150 ml/min). Results showed excellent VOC mass recovery (100% at flow rates ≤100 ml/min, and using 8 ml of solvent) and the $\delta^{13}C$ measurements conducted on TCE and benzene were identical to the source signature. Afterwards, the solvent-based dissolution method was used in a field study to sample gas-phase chlorinated solvents collected in the vadose zone overlaying a contaminated groundwater.[17] The transfer of gas-phase PCE and TCE into methanol was achieved by equilibrating 300 ml of sampled gas-phase with 2 ml of methanol in a closed vial. Nevertheless, the solvent method has not yet been tested to process air volumes by continuous pumping under field conditions.

The present study consisted in designing a preliminary dissolution-tube sampling device that will serve as a basis for future device development and to test the functionality under field conditions with the emphasis on TCE as the compound of interested. More specifically, the objectives were to investigate the reliability of the dissolution-tube sampling device in (i) collecting gas-phase TCE, (ii) providing reliable carbon and chlorine isotope results by comparing the results to a reference sampling technique (Summa canisters) and (iii) to link the isotope signature measured for gas-phase TCE to the signature of the emitting source.

## 2. METHODOLOGY

### 2.1 Air-solvent partitioning coefficient

Quantification of the air-solvent partitioning coefficient for TCE and benzene was performed in 40 ml glass vials. Ten (10) ml of methanol was injected followed by either 5 ul, 10 ul or 20 ul of TCE (or benzene) and the vial was then closed with Mininert caps. Each combination was reproduced in triplicate. Vials were gently shaken (200 rpm) and allowed to equilibrate over night at 25 °C. Sample of the gas-phase was taken using gas-tight syringe and analyzed for VOC concentration. Upon completion of gas-phase monitoring, the liquid phase was sampled and analyzed for VOC concentration.

### 2.2 Sampling device and flow rate

The proposed preliminary sampling device is illustrated on Figure 1. A 40 ml glass vial commonly used for groundwater sampling is filled with 30 ml of solvent, and connected to a pump via a metal adaptor. The metal adaptor was specially designed to avoid contact of the sampling gas with the pump. The adaptor consisted of two metal tubes, one extended to the bottom of the vial and the other to the headspace only. When pumping air through the shorter tube, a tension is created in the headspace of the vial, and hence the ambient air is pulled into the vial via the longer tube. The air migrates upward through the solvent as bubbles, and during the short transition time, the gas-phase VOC dissolves in the solvent. A sorbent filter (filled with activated carbon) connected along the outlet line allows to catch the solvent volatilized during the process. The filter can be as long as required to

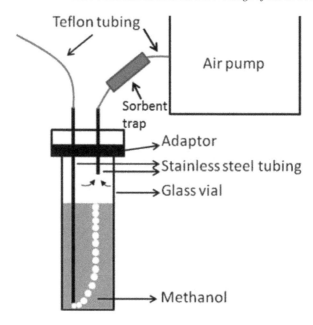

**Figure 1** *Schematic representation of the dissolution tube sampling device*

ensure no solvent breakthrough, as no analysis is meant to be performed on this sorbent matrix. The flow rate in presence of 30 ml of methanol was first assessed in the laboratory to insure a direct relationship between the imposed flow rate (by the pump) and the effective flow rate. An in-house gas pump was used and connected to the device as illustrated on Figure 1. The outlet of the pump was connected to a flow meter (red-y®, Vögtlin), hence measuring positive flow. The pump was sequentially set to 50 ml/min and 100 ml/min during 2 minutes and flow rate measurements were noted every 10 seconds.

## 2.3 Field site experiment

The study was conducted in a former industrial building in which an artificial TCE gas-phase plume was created in a selected room. The dimension of the room was 5 m wide, 7 m long, and 4.3 m high, for a total volume of 150 m$^3$ (Figure 2). A control sample of the indoor air was taken using a summa canister to assess the indoor air quality prior the experiment. A total of 500 ml of liquid phase TCE (VWR, 95% purity) was poured into a 2 L beaker and then placed and secured on the floor, against the wall. A time period of 67 h was allowed for the gas-phase TCE plume to form inside the room before performing the sampling events. The NAPL source remained in place throughout the course of the experiment, for a total duration of 92 h, hence insuring constant gas-phase TCE input in the room. During emplacement of the source and upon completion of the experiment, 1 ml of the NAPL was sampled (in duplicate) and injected into a 40 ml vial pre-filled with deionized water. The water samples containing the NAPL were sent to two laboratories in order to independently determine the initial and final $\delta^{13}C$ composition of TCE. However, the $\delta^{37}Cl$ analysis was conducted in only one laboratory (University of Neuchatel). Finally, the source weight was periodically recorded to evaluate the TCE mass lost during the course of the experiment.

Six 6 L summa canisters were divided in two groups, and each group was positioned in the room relative to the location of the source. Three canisters identified C-S0, C-S6 and C-S24 were placed 2 m away from the source (Figure 2). The intake rates of the canisters were respectively instant, 6 h and 24 h. Three additional canisters identified C-D0, C-D6 and C-D24 were placed approximately 8 m away from the source, at the opposite corner of the room, and pre-set with the same three intake rates as the front row group. Although the six canisters were simultaneously activated for sampling at t=67 h, only the canisters set for instant rates correspond to sampling event 1. The end of the sampling period for the canisters set for 6 h and 24 h correspond respectively to sampling events 2 and 3. The solvent-based dissolution method was used in parallel to the summa canisters during each sampling event and 30 ml of methanol was used as the sink matrix. Three sampling beginning moments coincided with the end of the respective intake rate of the canister, producing samples Ma (at t=67 h, sampling event 1), Mb (at t=73 h, sampling event 2) and Mc (at t=91 h, sampling event 3). The air from the middle of the room, in between the two canister rows, was pumped at a rate of 50 ml/min using a Gilian Sensidyne BDX II pump with a low flow rate adaptor (and previously calibrated using a bios dry DCL-ML device). For Ma sampling, 1.25 L of air was pumped. Sampling volume for Mb and Mc was arbitrarily increased to 3.0 L, to ensure sufficient dissolved mass required for analytical measurements. Upon completion of the sampling, the vial was filled with deionized water to avoid any headspace and then stored at 4 °C. Finally, a photoionization detector (PID) (Minirae 2000) was placed in the center of the room (Figure 2) and was set to log measurements every minute starting from the source emplacement event. The PID was previously calibrated with isobutylene and the given correction factor to convert the readings to TCE concentration was used.

## 2.4 Analytical method

*2.4.1 Concentration analysis*
Concentration analysis for summa canisters was initiated with pressurization to 25 psia. Then, 200 ml of gas was transferred into a concentrator (7100 Concentrator; Entech Instruments, Simi Valley) to prepare the sample for GC analysis. The concentrator removed water, $CO_2$, $N_2$ and $O_2$ and cryofocused the VOCs. The VOCs were then thermally desorbed into a stream of $N_2$ and passed into a gas chromatograph (GC; Trace; Fisher Scientific Pittsburgh, PA). The VOCs were separated on RTx1 column (Restek Inc, Bellefonte, PA) and the effluent was passed into a quadrupole mass spectrometer (DSQ, Fisher Pittsburgh, PA) for positive identification and quantification. Relative standard deviation of reference standards was less than or equal to 10% for five calibration standards.

*2.4.2 $\delta^{13}C$ measurements.* To perform $\delta^{13}C$ analysis for TCE dissolved in liquid (water or methanol), a sample aliquot was diluted in 40 ml glass vial filled with deionized water. Dilutions were calculated based on concentration results obtained with the canisters in order to inject approximately 10 nmol of carbon into the IRMS detector. The diluted samples were analyzed using a purge and trap (P&T) module (Tekmar Velocity, USA) connected to a TRACE gas chromatograph (GC) coupled to a ThermoFinnigan Delta Plus XP IRMS via a ThermoFinnigan GC combustion III interface. Twenty five (25) ml of the sample was purged and the extracted VOC was trapped on a Vocarb 3000 sorbent matrix (Supelco). After the 10 minute purging step, the gas-phase VOC mass was transferred to a cryogenic trap (Atas, Nederlands) to concentrate the analyte into a narrow band prior

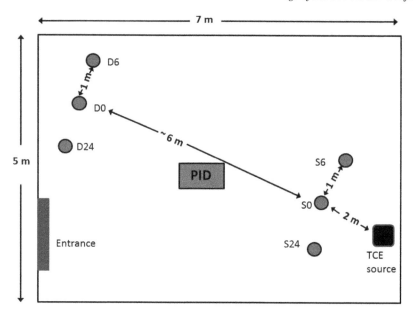

**Figure 2** *Emplacement of the summa canisters, PID and the TCE source inside the room*

injection to the column. A DB-VRX column (BGB Analytik, Bockten, Switzerland, 60 m × 0.25 mm i.d.) was used and the carrier gas was helium at a flow rate of 1.7 mL/min. The oven initial temperature was kept at 50 °C for 10 minutes and increased to 120 °C at a rate of 10 °C/min. Standard deviation of reference standards made with 1% of methanol in water (V:V) was 0.41‰ (n= 5). The $\delta^{13}C$ results measured by the two laboratories for the same TCE material were noted to differ by 0.8‰. The variation is likely due to different IRMS calibrations. Hence, a correction was applied to the data to override this inter lab analytical difference. For $\delta^{13}C$ analysis with summa canister, the pressurized canister was attached to an autosampler (3100, Entech Instruments, Simi Valley) which directed the sample into a concentrator as described earlier. The amount of volume transferred was based on concentration results to deliver 10 nmol of carbon into the IRMS detector. Following the preparation step (gas removal and VOC cryofocusing), the VOCs were transferred into a GC (Ultra-Trace II, Thermo Fisher Scientific, Pittsburgh, PA) connected to an Isotope Ratio Mass Spectrometer (IRMS Delta V, Thermo Fisher Scientific, Pittsburgh, PA) via a combustion interface (GCC-III, Thermo Fisher Scientific, Pittsburgh, PA). The carrier gas was helium at a flow rate of 1.2 mL/min and a RTx1 column (Restek Inc, Bellefonte, PA) was used. Carbon isotope ratios are reported using the $\delta^{13}C$ notation (in ‰) relative to the VPDB standard according to the following:

$$\delta = (R/Rstd - 1)*1000 \; [‰] \tag{1}$$

Where R and Rstd are respectively the isotope ratio of the sample and the international reference standard (VPDB for carbon).

$\delta^{37}Cl$ measurements: For water and methanol samples, an aliquot was diluted in 15 ml of deionized water using 20 ml glass vials to obtain approximately 150 µg/L of TCE. The diluted samples were then analyzed with an Agilent 7890A gas chromatograph (GC) coupled to an Agilent 5975C quadrupole mass selective detector (qMS) (Santa Clara, CA,

USA). One thousand (1000) μl of the headspace was injected using a CombiPal Autosampler (CTC Analytics, Zwingen, Switzerland). A DB-5 column (30 m × 0.25 mm i.d., Agilent) was used and helium served as a gas carrier (1.2 mL/min). The oven initial temperature was held at 40 °C for 2 minutes and increased to 85 °C at a rate of 15 °C/min and further increased to 150 °C at a rate of 30 °C/min. Each vial was analyzed five times. Chlorine isotope ratios were determined using a method proposed by Sakaguchi-Soder, *et al.* [18] and modified by Aeppli, *et al.*[19] The latter method quantifies the four chlorine isotopologues of TCE (mass 130, 132, 134, and 136), which are then used to determine the raw isotopic ratios based on the following equation (2):

$$R\_TCE = (i\_132 + [2*i]\_134 + [3*i]\_136) / ([3*i]\_130 + [2*i]\_132 + i\_134) \quad (2)$$

Where i corresponds to the molecular ion abundance at different m/z values.

Values obtained with equation 2 were compared to a calibration curve established with two isotopically different reference standards (3.05‰ and −2.70‰) to derive delta values on the SMOC scale, as recommended by Bernstein, *et al.*[20] Standard replicates were analyzed at the beginning and at the end of each analytical sequence and the standard deviation (n=7) was 0.48‰. No $\delta^{37}Cl$ measurement was performed on the samples taken with the summa canisters.

## 3. RESULTS AND DISCUSSION

### 3.1 Laboratory experiments

*3.1.1 Air-solvent partitioning coefficient.* While the pumped air is continuously bubbling through the solvent during the sampling process, the VOC is expected to dissolve in the solvent until the concentration equilibrium between the gas-phase and the organic solvent is reached. Therefore, and contrasting with sorbent tubes, the maximum concentration possible in the solvent hence depends on the gas-phase concentration and not on the volume of air pumped. The affinity of the targeted contaminant with the solvent hence becomes a critical parameter and can be quantified upon determination of the air-solvent partitioning coefficient. The partitioning coefficient represents the ratio of gas-phase to liquid-phase VOC concentration (unit less). Accordingly, TCE and benzene partitioning coefficients were experimentally measured for methanol and were respectively 0.0018 ±0.0006 and 0.0012 ±0.0001, indicating a slightly better affinity for benzene with methanol compared to TCE. The value measured for benzene is in good agreement with the calculated value of 0.0016 reported by Abraham, *et al.*[21] Although no value is reported for TCE, our value measured is somehow comparable to 1,1 dichloroethane (1,1 DCA) reported to be 0.0069,[21] still indicating lesser affinity for chlorinated compounds to methanol compared to benzene.

Using the measured air-solvent partitioning coefficients, the analytical restriction covered in [16] (maximum solvent load of 1% in water) and owing a minimum of 1 nmol of carbon delivered to the ion source, the minimal VOC concentration in the air in order to accumulate enough carbon mass for reliable isotope measurements refers to the detection limit of the method. Accordingly, the detection limit for reliable TCE and benzene assessment is respectively estimated to 480 +/- 160 μg/m$^3$ and 65+/- 5 μg/m$^3$. For benzene, although an air-solvent partitioning coefficient in the same range than TCE, the detection limit is significantly lower and is due to the presence of more carbon in the benzene molecular structure, hence requiring less mass to achieve 1 nmol of carbon.

**Table 1** *Flow rate variations related to the negative pressure created during sampling with the dissolution tube device. The test was performed twice, where flow measurement was taken every 10 seconds over 2 minutes*

| Flow rate set point (ml/min) | Sampling device setup | Average reading *** (ml/min) | std dev (ml/min) |
|---|---|---|---|
| 50 | without* | 50.0 | 3.5 |
|    | with**   | 46.0 | 5.1 |
| 100 | without* | 104.5 | 6.8 |
|     | with**   | 106.0 | 5.5 |

\* Pumping air without methanol
\*\* Pumping air with presence of methanol
\*\*\* Average of 2 replicates

*3.1.2 Flow rate measurements.* Accurate knowledge of the gas volume circulating through the solvent is key to determine the pumping time and keeping the sampling event in a constrained timeframe. Furthermore, when gas-phase concentrations at the field site are known, the minimal required air volume to load the solvent with sufficient VOC mass or to saturate the solvent can be accurately determined. Table 1 shows the variation of the flow rate when a tension is created in the headspace of the vial in presence of methanol. The results indicated negligible variations in flow rates whether or not methanol was used, hence indicating that the volume of air pumped can be accurately determined despite the negative pressure required to circulate the air.

### 3.2 Field experiment

*3.2.1 TCE source monitoring.* The initial and final weight of the TCE source were respectively 1128 g and 1077 g, for a total vaporization of 51 g of TCE over 9 2h. The small proportion of mass loss (4.5%) implies a TCE input in the air throughout the course of the experiment, hence avoiding gas-phase concentration decrease due to source attenuation. The initial and final $\delta^{13}C$ and $\delta^{37}Cl$ values for TCE at the source are illustrated in Figure 3a. When comparing the initial to the final $\delta^{13}C$ values, the results indicate stability of the source isotopic signature for both elements throughout the course of the experiment. Although a recent study showed that progressive vaporization of TCE NAPL leads to a depletion in $\delta^{13}C$ and to an enrichment in $\delta^{37}Cl$,[22] the small TCE mass vaporized during the span of the experiment was too low to create a significant isotopic shift.

*3.2.2 TCE concentration in indoor air.* Both the control sample taken with the canister and the PID measurements recorded before source placement showed non-detect results, confirming a TCE-free indoor air prior the experiment. The concentrations of TCE in indoor air following the source placement measured using the summa canisters during the three sampling events are listed in Table 2. In addition, Table 2 provides the average PID measurements covering the same intake time period as the summa canisters and Figure 4 shows the concentration evolution monitored by the PID during the course of the experiment. The PID measurement equivalent to the canister instant intake rate is an average reading of 10 minutes. Except for the concentration spike of 12.9 ppmv observed soon after source emplacement, continuous measurements with the PID were relatively

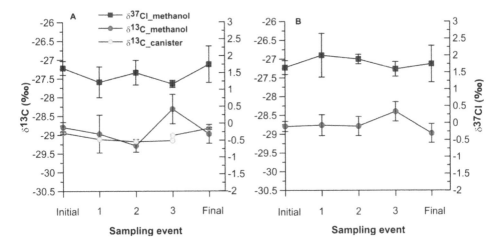

**Figure 3** $\delta^{13}C$ and $\delta^{37}Cl$ for TCE at the source (initial and final sampling) and sampled in the gas-phase using canisters and dissolution tubes (methanol) for three sampling events. Figure A: analysis carried out in the lab <1 month after sampling event 3. Sampling event 1 includes: C_D0 and Ma. Sampling 2 includes: C_S6 and Mb. Sampling event 3 includes: C_S24, C_D24 and Mc. Figure B: analysis carried out in the lab ~365 days after the sampling event 3. Error bars represent the standard deviation of 3 and 5 analyses for $\delta^{13}C$ and $\delta^{37}Cl$, respectively

stable over time and varied between 0.9 and 1.7 ppmv during the time period covering the three sampling events. Note that a problem with the PID occurred which impeded monitoring the entire duration of the experiment. Concentration results monitored with the summa canisters show larger variation in concentration compared to the PID over time and space in the room, and varied between 0.820 and 4.17 ppmv. In addition, lower concentrations were observed in 2 out of 3 canisters placed near the source compared to the canisters placed further away with equivalent intake rate. Increasing concentrations with increasing distance from the source suggests a non-uniform gas-phase TCE distribution in the room. These counterintuitive results nevertheless agree with previous studies where VOC concentrations in indoor air were shown to be spatially variable at very small scale due to poor air mixing regulated by diffusion and convection.[23,24]

**Table 2** *Indoor air TCE concentrations measured with the canisters and the PID for different time periods and location inside the room*

| Intake rate | Canister | | PID | |
|---|---|---|---|---|
| | Conc near | Conc away | Conc | std dev |
| | ppmv | ppmv | ppmv | ppmv |
| Instant | 0.82 | 1.75 | 1.2 | 0.1 |
| 6 h | 2.29 | 3.71 | 1.5 | 0.3 |
| 24 h | 4.17 | 2.59 | 1.7 | 0.4 |

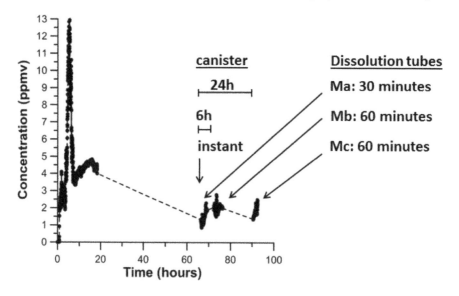

**Figure 4** *TCE concentration recorded by the PID over the course of the experiment. The sampling moment for each device is approximately indicated relatively to the time course of the experiment. The length of the line below the 6 h and 24 h canister represents the sampling time period covered*

### 3.3 $\delta^{13}$C measured with summa canisters

The $\delta^{13}$C values measured for TCE sampled with the canisters set at three different intake rates and at two different distances from the source are presented in Figure 3a. Globally, the $\delta^{13}$C value measured with every canister showed no difference when compared to the source signature. The similarity in $\delta^{13}$C values measured over the three sampling events underlines two important facts regarding sampling with summa canisters. First, the intake rate has no effect on the $\delta^{13}$C values over the range tested (instant, 6 h, 24 h). This is supported by comparing results from C-S6 to C-S24 and C-D0 to C-D24. These results suggest the possibility of sampling with summa canisters at any intake rate without significantly affecting the true $\delta^{13}$C value of the TCE in the indoor air. Second, with respect to the concentration variation magnitude observed in the building within a 24 h long monitoring, the $\delta^{13}$C composition of the gas-phase TCE did not change significantly. This is supported by comparing results from C-S24 to C-D24. The results hence suggest that CSIA sampling does not need to be performed at specific place inside the breathing space of the room. However, due to mechanical problems during the analysis, no $\delta^{13}$C result can be provided for samples C-S0 and C-D6 which significantly limits a detailed spatial assessment.

### 3.4 $\delta^{13}$C and $\delta^{37}$Cl measured with dissolution tubes

The $\delta^{13}$C and $\delta^{37}$Cl values for TCE sampled using the dissolution tubes over three sampling events are also presented in Figure 3a. Based on the acceptable deviation limit of

0.5‰, the $\delta^{13}C$ and $\delta^{37}Cl$ values measured for samples Ma, Mb and Mc showed no significant difference when compared to the source signature. The variation of the $\delta^{13}C$ values is most likely not related to TCE breakthrough. The pumping rate of 50 ml/min used in this study was selected as Bouchard and Hunkeler [16] reached the expected TCE mass recovery while injecting TCE-loaded air at this rate through only 8 ml of methanol. In addition, the same study showed no influence of the mass loss on the $\delta^{13}C$ value when using higher flow rates (100 and 150 ml/min). Observed variations shown by the error bars in Figure 3a are more likely due to the injection of a large mass of methanol which requires special attention during the analytical measurement. The $\delta^{37}Cl$ results are also in agreement with the source signal. The $\delta^{37}Cl$ values are within the acceptable deviation limit of 0.5‰. Furthermore, no temporal variation was simultaneously observed for the $\delta^{13}C$ and $\delta^{37}Cl$ values during short term sampling, which is in agreement with the stability observed for $\delta^{13}C$ measured with the summa canisters over shorter and longer periods of time. Although no significant change in the $\delta^{13}C$ TCE composition was observed in this study over time and space, a more detailed investigation nevertheless remains to be conducted at lower indoor air concentration to fully demonstrate spatial isotopic signature stability over a longer period of time in a larger room and also among several interconnected rooms.

**3.5 Sample preservation**

To investigate the preservation capacity of the solvent, the field samples were stored at 4°C and the $\delta^{13}C$ and $\delta^{37}Cl$ analyses were once more performed approximately 1 year later for Ma, Mb and Mc. The results are presented in Figure 3b. The values showed no significant change compared to the values that were obtained when the samples were first analyzed within 1 month after sampling. The results confirm the excellent preservation capacity offered by the solvent as it provides an excellent sink medium for VOCs and prevents their degradation. The good preservation of the sample over time hence reduces the need for a rapid turnaround and allows analyzing supplemental solute afterwards.

4. SUMMARY

This work presented a novel field sampling methodology that consists on dissolving gas-phase VOC into solvent while sampling in view to perform CSIA. To test the preliminary sampling device, a comparative sampling was conducted during a field experiment with TCE as compound of interest. Both sampling devices (summa canisters and dissolution tubes) showed similar $\delta^{13}C$ results, and the values obtained for both elements ($\delta^{13}C$ and, $\delta^{37}Cl$) were in agreement with the known isotopic composition of the TCE source. In addition, the field experiment also allowed underlining interesting characteristics regarding $\delta^{13}C$ and $\delta^{37}Cl$ for TCE monitored in indoor air. The results obtained in our experimental setup showed no significant isotopic fractionation related to short temporal and spatial concentration variations. According to our 24 h sampling timeframe, the isotopic results can be used to establish a link between gas-phase VOC monitored in indoor air and an emanating indoor source. Hence, for sites with multi potential sources, CSIA will contribute at indicating the true emitting source. The absence of isotope fractionation would significantly simplify the use of CSIA method to conduct forensics investigation, but this stability remains to be demonstrated for buildings with several rooms interconnected and subjected to large concentration gradients.

The simplicity of the sampling device and the good isotope reproducibility demonstrated during the field scale experiment hence offer an alternative sampling method to conventional sorption tubes to perform CSIA in gas-phase VOCs studies. Assets of the solvent-based method include the possibility to perform dual isotopic signature assessment ($\delta^{13}C$ and $\delta^{37}Cl$, and eventually $\delta^2H$) on the same sample, to conduct duplicate analysis for each sample, and the excellent preservation capacity of the methanol sample over time (over 1 year). Further tests on the method should include the evaluation of the sampling methodology under very low gas-phase concentration (where larger air volumes need to be processed), determining the most suited solvent, and evaluating post-sampling concentrating procedures. Should the concentrating step be effective, the sampling methodology will highly benefit by decreasing both the sampling time and the detection limit.

**Acknowledgments**

The authors are grateful to DuPont (James K. Henderson) for providing the funding required to perform the field experiment and to Luc Boisseau (SNC Lavalin Environment, inc.) for carrying out the field work.

**References:**

1. Hunkeler, D.; Chollet, N.; Pittet, X.; Aravena, R.; Cherry, J. A.; Parker, B. L., Effect of source variability and transport processes on carbon isotope ratios of TCE and PCE in two sandy aquifers. J. Contam. Hydrol. 2004, 74, (1-4), 265-282.
2. Blessing, M.; Schmidt, T. C.; Dinkel, R.; Haderlein, S. B., Delineation of Multiple Chlorinated Ethene Sources in an Industrialized Area-A Forensic Field Study Using. Environ. Sci. Technol. 2009, 43, (8), 2701-2707.
3. Aelion, C. M.; Hoehener, P.; Hunkeler, D.; Aravena, R., Environmental isotopes in biodegradation and bioremediation. CRC Press. Taylor and Francis Group: Boca Raton, London, New York, 2010; p 450.
4. Pond, K. L.; Huang, Y. S.; Wang, Y.; Kulpa, C. F., Hydrogen isotopic composition of individual n-alkanes as an intrinsic tracer for bioremediation and source identification of petroleum contamination. Environ. Sci. Technol. 2002, 36, (4), 724-728.
5. Mansuy, L.; Philp, R. P.; Allen, J., Source identification of oil spills based on the isotopic composition of individual components in weathered oil samples. Environ. Sci. Technol. 1997, 31, (12), 3417-3425.
6. Turner, N.; Jones, M.; Grice, K.; Dawson, D.; Ioppolo-Armanios, M.; Fisher, S. J., delta C-13 of volatile organic compounds (VOCS) in airborne samples by thermal desorption-gas chromatography-isotope ratio-mass spectrometry (TD-GC-IR-MS). Atmospheric Environment 2006, 40, (18), 3381-3388.
7. Vitzthum von Eckstaedt, C.; Grice, K.; Ioppolo-Armanios, M.; Jones, M., delta C-13 and delta D of volatile organic compounds in an alumina industry stack emission. Atmospheric Environment 2011, 45, (31), 5477-5483.
8. Vitzthum von Eckstaedt, C.; Grice, K.; Ioppolo-Armanios, M.; Chidlow, G.; Jones, M., delta D and delta C-13 analyses of atmospheric volatile organic compounds by thermal desorption gas chromatography isotope ratio mass spectrometry. Journal of Chromatography A 2011, 1218, (37), 6511-6517.
9. Goldstein, A. H.; Shaw, S. L., Isotopes of volatile organic compounds: An emerging approach for studying atmospheric budgets and chemistry. Chem. Rev. 2003, 103, (12), 5025-5048.

10. Tsunogai, U.; Yoshida, N.; Gamo, T., Carbon isotopic compositions of C-2-C-5 hydrocarbons and methyl chloride in urban, coastal, and maritime atmospheres over the western North Pacific. J. Geophys. Res.-Atmos. 1999, 104, (D13), 16033-16039.
11. McHugh, T.; Kuder, T.; Fiorenza, S.; Gorder, K.; Dettenmaier, E.; Philp, P., Application of CSIA to Distinguish Between Vapor Intrusion and Indoor Sources of VOCs. Environ. Sci. Technol. 2011, 45, (14), 5952-5958.
12. Zhang, L.; Bai, Z.; You, Y.; Wu, J.; Feng, Y.; Zhu, T., Chemical and stable carbon isotopic characterization for PAHs in aerosol emitted from two indoor sources. Chemosphere 2009, 75, (4), 453-461.
13. Shouakar-Stash, O.; Frape, S. K.; Drimmie, R. J., Stable hydrogen, carbon and chlorine isotope measurements of selected chlorinated organic solvents. J. Contam. Hydrol. 2003, 60, (3-4), 211-228.
14. Jia, C.; Batterman, S.; Godwin, C., VOCs in industrial, urban and suburban neighborhoods, Part 1: Indoor and outdoor concentrations, variation, and risk drivers. Atmospheric Environment 2008, 42 2083–2100.
15. Klisch, M.; Kuder, T.; Philp, R. P.; McHugh, T. E., Validation of adsorbents for sample preconcentration in compound-specific isotope analysis of common vapor intrusion pollutants. Journal of Chromatography A 2012, 1270, 20-27.
16. Bouchard, D.; Hunkeler, D., Solvent-based dissolution method to sample gas-phase volatile organic compounds for Compound-Specific Isotope Analysis. Journal of Chromatography A 2014, 1325, 16-22.
17. Patterson, B. M.; Aravena, R., Davis, G. B.; Furness, A. J.; T.P., B.; Bouchard, D., Multiple lines of evidence to demonstrate vinyl chloride aerobic biodegradation in the vadose zone, and factors controlling rates. J. Contam. Hydrol. 2013, 153, 69–77.
18. Sakaguchi-Soder, K.; Jager, J.; Grund, H.; Matthaus, F.; Schuth, C., Monitoring and evaluation of dechlorination processes using compound-specific chlorine isotope analysis. Rapid Commun. Mass Spectrom. 2007, 21, (18), 3077-3084.
19. Aeppli, C.; Holmstrand, H.; Andersson, P.; Gustafsson, O., Direct Compound-Specific Stable Chlorine Isotope Analysis of Organic Compounds with Quadrupole GC/MS Using Standard Isotope Bracketing. Anal. Chem. 2010, 82, (1), 420-426.
20. Bernstein, A.; Shouakar-Stash, O.; Ebert, K.; Laskov, C.; Hunkeler, D.; Jeannottat, S.; Sakaguchi-Soder, K.; Laaks, J.; Jochmann, M. A.; Cretnik, S.; Jager, J.; Haderlein, S. B.; Schmidt, T. C.; Aravena, R.; Elsner, M., Compound-Specific Chlorine Isotope Analysis: A Comparison of Gas Chromatography/Isotope Ratio Mass Spectrometry and Gas Chromatography/Quadrupole Mass Spectrometry Methods in an Interlaboratory Study. Anal. Chem. 2011, 83, (20), 7624-7634.
21. Abraham, M. H.; Whiting, G. S.; Carr, P. W.; Ouyang, H., Hydrogen bonding. Part 45. The solubility of gases and vapours in methanol at 298 K: An LFER analysis. Journal of the Chemical Society-Perkin Transactions 2 1998, (6), 1385-1390.
22. Jeannottat, S.; Hunkeler, D., Chlorine and Carbon Isotopes Fractionation during Volatilization and Diffusive Transport of Trichloroethene in the Unsaturated Zone. Environ. Sci. Technol. 2012, 46, (6), 3169-3176.
23. McClennen, W. H.; Vaughn, C. L.; Cole, P. A.; Sheya, S. N.; Wager, D. J.; Mott, T. J.; Dworzanski, J. P.; Arnold, N. S.; Meuzelaar, H. L. C., Roving GC/MS: Mapping VOC gradients and trends in space and time. Field Analytical Chemistry and Technology 1996, 1, (2), 109-116.
24. Kim, S. K.; Burris, D. R.; Bryant-Genevier, J.; Gorder, K. A.; Dettenmaier, E. M.; Zellers, E. T., Microfabricated Gas Chromatograph for On-Site Determinations of TCE in Indoor Air Arising from Vapor Intrusion. 2. Spatial/Temporal Monitoring. Environ. Sci. Technol. 2012, 46, (11), 6073-6080.

# CHARACTERISATION AND MODELLING OF DUST IN A SEMI-ARID CONSTRUCTION ENVIRONMENT

John Bruce[1,2], Hugh Datson[2], Jim Smith[1] and Mike Fowler[1]

[1]Earth and Environmental Sciences, University of Portsmouth, Burnaby Building, Burnaby Road, Portsmouth PO1 3QL, UK
[2]DustScan Ltd., Griffin House, Market Street, Charlbury, Oxford OX7 3PJ, UK

## 1 INTRODUCTION

Dust from a wide range of industrial activities has the capacity to cause environmental concerns. Dust may be generated in a variety of ways, from both construction and subsequent industrial activities, and may contain unwanted additions from the relevant activities and processes. The impacts may therefore vary from general nuisance to health implications, based on the size, volume and content of the dust.

'Dust' is defined by BS 6069 Part 2[1] as particulate matter <75μm in diameter. It can be split into two general size fractions: coarse particles (>10μm) perceived with annoyance risk and finer particles associated with risk to human health. Finer size fractions of dust are clearly defined and are used to indicate air quality through the Air Quality Objectives (AQO[2]) and National Air Quality Strategy (NAQS[2]). Coarse particles however are poorly defined and are often referred to as 'nuisance dust'. The 'nuisance' caused is mostly derived from the chronic soiling of surfaces from deposited dust, or from acute problems such as passing dust clouds. Assessment of nuisance dust is through criteria developed by the Institute of Air Quality Management (IAQM[3]), National Planning Policy Framework (NPPF[4]) and the Mineral Industry Research Organisation (MIRO[5]).

Nuisance dust modelling is a complex field which is complicated by the potential for 'unlimited' sources. In standard theoretical modelling, any activity that emits a significant emission must be designated with an emission 'factor', in order to predict how it might disperse. For dust emissions, any activity that results in the entrainment of dust can be deemed an emission: this may include wind erosion of stockpiles or bare ground, site movements and mechanical processes. Due to the complexity of measuring these emissions individually, coarse dust modelling emission factors do not exist[6] and are often ignored. A simple technique to predict future dust movements may be to model future levels based on previous observations. This is explored in a case study that investigates the characteristics and modelling of dust. It is imperative to understand both what is contained in the dust and what has caused it so that future movements can be understood and predicted.

## 2 STUDY SITE

The study site is located on the coast of the Caspian Sea, in an area where natural dust is an issue of environmental concern due to the semi-arid climate with strong prevalent winds from the north. The main working site is a large oil and gas terminal where dust movements were first studied using monitoring undertaken as part of an environmental and socio-economic impact assessment (ESIA) for proposed construction work. Two periods of dust monitoring were undertaken: initial monitoring over three months as part of a baseline study, and further monitoring for 18 months during a preliminary infrastructure construction phase. Comparisons between the two were made to evaluate the impact of the construction workings on dust in the area. A modelling study was undertaken to help understand what caused the dust trends observed, why they occurred during certain weather conditions and ultimately predict dust patterns generated by natural sources. This would enable the site operator to understand any dust impacts and what the underlying causes were. Weather data were also collected at an onsite meteorological station.

## 3 DUST MONITORING

Passive dust sampling was undertaken at ten initial monitoring stations, expanding to fourteen for the main investigative period, each with a DustScan directional and depositional sticky pad dust collector. Monitoring locations (Figure 1) included an array of background samplers (upwind), samples at site boundaries, downwind samples and samples taken at nearby population centres (receptors). The array enabled potential pathways of fugitive dust from sources to receptors to be investigated and understood.

Sticky pad monitors are routinely exposed for periods of a few days up to a recommended maximum of two weeks and are sealed with a transparent film upon completion[7]. Samples are scanned and analysed using bespoke software where measurements for dust coverage (Absolute Area Coverage – AAC %) and dust soiling (Effective Area Coverage – EAC %) are made to give an indicative risk value for dust nuisance. The sticky pads have an adhesive that can be decoupled to enable the encapsulated dust to be removed and filtered for further analysis[7]. With directional dust samples analysed in 15° segments, sub sampling can be undertaken based on the inferred point of origin and subsequent pathway.

## 4. DUST CHARACTERISATION STUDY

### 4.1 Analytical methods

A comprehensive suite of analytical methods was used to test selected dust samples from both the baseline study and the main monitoring period. Soil samples at dust monitoring points, road dust samples and grab samples from construction spoil heaps were also tested for comparison. Sticky pad sub-samples were taken from both directional and depositional dusts from receptors, background locations and on site boundaries. Sub sampling from directional dust samples used 15° arcs of interest from the relevant direction. The characterisation study was undertaken using mass spectrometry, mineral analysis and particle size grading. This combination of analyses allowed for changes in dust content to be comprehensively assessed.

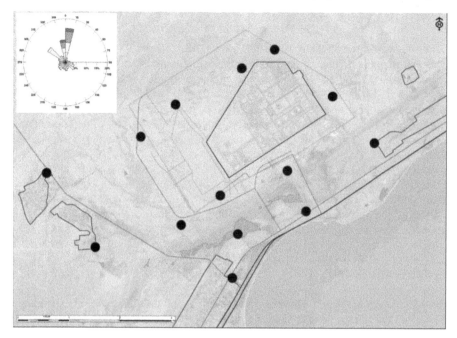

**Figure 1** *The location of fourteen dust monitoring stations (black dots). The central green line outlines the site ownership boundary, the black box is the current site and the orange box is the construction expansion area. The south-eastern green line is a main road where road dust samples were also taken. The blue line across the coast is a motorway. Receptors can be found in the south, east and west, and are denoted by red boxes.*

Elemental analysis was undertaken using an Inductively Coupled Plasma Mass Spectrometer (ICP-MS) at the University of Portsmouth. The model used was an Agilent 7500ce, and sample preparation was completed as explained in Fowler et al[8]. Particle size analysis was determined by laser granulometry using a Malvern Mastersizer. Mineral characterisation was undertaken by QEMSCAN, an automated mineral analysis system using Scanning Electron Microscopy - Energy Dispersive X-Ray Spectrometry (SEM-EDX). Sticky pad directional and depositional dust samples, soil samples, road dust samples and spoil heap samples were analysed using all three methods.

## 1.2 ICP-MS Results

The results of ICP-MS analysis are summarised in Figure 2, showing the proportions of major elements in a selection of samples. Silicon is not included because it is lost as a vapour during sample preparation.

Figure 2 shows that the element proportions were similar for both dust samples and soil samples, although it can be noted that there was considerably more Ca in the road dust samples than the other samples. There was also an increased proportion of Ca in the southern receptor dust samples; indicating a higher proportion of calcitic material in the dust arising from road traffic or local construction activities.

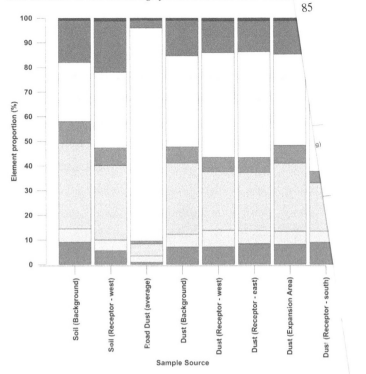

**Figure 2** *Major elemental proportion results from ICP-MS analysis* dust samples.

Excluding the road dust samples, Na and K were at reasonably [
most samples comprising 10% of the dust and soil. Other elements su[
widely; this may be linked to mineralogy. Correlations between b[
background dusts were strong ($r^2=0.79$) and dusts from across the site
($r^2=1$).

### 4.3 QEMSCAN Results

QEMSCAN results are summarised in Figure 3 with the major mi[
reported. The mineralogical characteristics of the dust and soil sam[
results, although good correlations were again found between backg
background dusts ($r^2=0.74$). Soil sample mineralogy was varied however, v
in receptor samples in addition to the road dust samples. This is consiste[
proportions of calcium found in the ICP-MS results. Most dust samples co
proportions of major minerals as the background soil, indicating that the d
derived from local soils.

The very high proportion of calcite in the road dust and higher proporti[
at receptor is considered to be due to spillages from road traffic. Nearby hill
for calcite-rich limestone, which is transported by road via local receptors, an
limestone from road transport are common. In summary, QEMSCAN a[
consistent with findings from the ICP-MS results; dusts were derived mostl
soils.

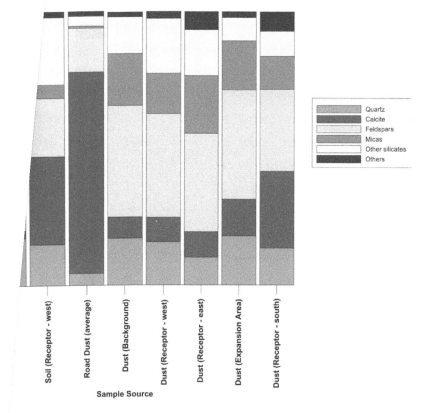

Major mineral proportion results from QEMSCAN analysis of dust, soil and road [dust sample]s.

### Granulometry

[Parti]cle size grading for dust samples are summarised in Figure 4, where dust particle sizes [at differ]ent locations during the same week ranged from less than 1μm to above 2000μm [in diame]ter. Over 50% of particles were less than 100μm, and there were no major [differenc]es in average particle size grading between dust localities. Figure 4 also shows an [emergin]g bi-modal distribution; this could be associated with different wind profiles over [the week]-long dust sampling period.

[Dust s]ample results were also similar across the site during other monitoring periods, [but vari]ed temporally in response to differing weather conditions. Higher proportions of [fine p]articles were observed during periods of high wind; this is thought to be due to the [increas]ed energy required to suspend large particles[9].

### Summary of results

[Overal]l, dusts and soils were found to have the same elemental and mineralogical [compo]sition with dusts during both the baseline and construction phases. Dusts from the [backg]round, expansion area and receptors were also comprised of similar major minerals

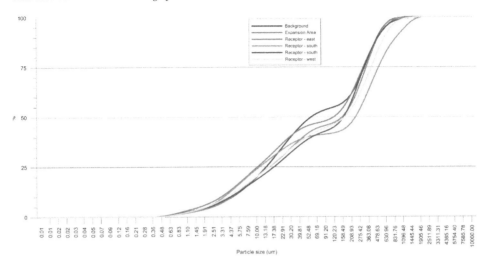

**Figure 4** *Particle size grading (cumulative) plot for dust samples from the same monitoring period.*

($r^2$ = 0.78) to the calcareous sandy silts and clays that make up the majority of local soils. One exception was in concentrations of calcite, which was increased in dust samples close to roads compared to dust samples from other areas. The only other difference in dust composition was in particle size grading, with grain sizes changing by week but maintaining similarities throughout the sample area for a given sample period.

## 5 DUST MODELLING STUDY

For this preliminary trial, the first twelve months of data from the main sampling period was compared with weather data collected from an onsite monitoring station. This firstly enabled the observed dust trends to be understood in terms of cause and effect. Patterns emerged indicating that background dust levels were closely linked to weather parameters. This included high dust levels originating from north of the site, which were observed to correlate with respective high winds from the north. Dust levels were also significantly increased during summer months and after periods of low rainfall. A study was undertaken in order to produce a basic empirical model, using weather parameters to predict future dust levels.

### 5.1 First steps

One key part of the study was to establish which collected dust was to be modelled and using what measurement. From initial comparisons of dust levels compared to weather, it was decided that 'background dust' would be modelled; this was defined as that arising from north of the site (330- 30°) and captured on the four most northern directional dust monitors. Both dust coverage (AAC%) and dust soiling (EAC%) measurements were analysed from background dust levels, but dust soiling of the sticky pads (EAC%) was chosen to be modelled as it had the advantage over dust coverage which tended to 'max-

**Table 1** *Weather parameters tested in the correlation matrix for rainfall only*

| Weather variable | Description |
| --- | --- |
| Total rain | Total rain over the monitoring period (MP) |
| Rain/day | Total rain over the MP, divided by number of days during the MP |
| Rain/ average | Average rain over the last 3 MP |
| Rain/day/factored | Rain/day factored over the last 3 MP, with increased influence for most recent (45/30/25%) |
| Rain risk | A coefficient generated by a lookup table for factored rain/day |
| Days of rain | Total days where $\geq 0.1$ of rain fell during the MP |
| Ratio days of rain | Days of rain divided by days during MP |
| Average days of rain | Average days of rain over the last 3 MP |
| Average days of rain/factored | Average days of rain over the last 3 MP, with increased influence for most recent (45/30/25%) |

out' at 100% during heavy periods of dusting. Furthermore, dust soiling levels were averaged over each monitoring period to give EAC%/day – thus taking into account the variability in sample monitoring period length. This allowed the simplified initial model to use only dust from natural causes, allowing for basic principles to be established before dusts could be attributed to specific site workings or processes. Deposited dust levels were also collected but not modelled due to uncertainties in the collection efficiency of sticky pads at high wind speeds.

### 5.2 Model development

A correlation matrix was created to establish the strongest links between background dust levels and four weather parameters: rainfall, temperature, wind speed, and wind direction. Each parameter was tested in a set of variables against background dust levels, with an example for rainfall shown in Table 1. The three variables with the best fit were chosen, with one from each parameter; average daily temperature over the monitoring period (MP), average daily rainfall over the MP and a 'wind risk' coefficient based on the average wind speed and the proportion of wind from the north during the MP. This coefficient was created to reflect the intensity and duration of winds from the north for each sample period.

Multiple linear regression was used to combine the three variables into a model. This attempts to model the relationship between two or more explanatory variables and a dependant variable by fitting a linear equation to the observed data. Each explanatory variable is given an individual constant by which it is multiplied. The three variables are then added together in combination with an intercept to create a best fit with the dust levels measured (Figure 5). An example of the data set for each variable is shown in Table 2. Temperature and wind risk had the biggest influence, with both demonstrating a positive correlation with dust movements, whilst rainfall had a smaller influence (approximately 10%) on dust predictions with more rainfall resulting in decreased modelled dust. This may be specific to the site; average daily temperature fluctuations range from of >30°C to <0°, which may cause temperature to having an increased effect on dust generation.

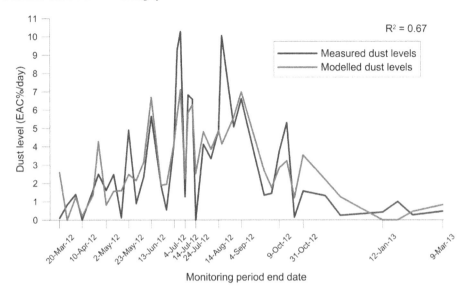

**Figure 5** *Measured dust levels plotted against modelled dust levels for the initial 12 month monitoring period.*

### 5.3 Model evaluation

The resulting model, based on the first year of sampling, fit the data well ($r^2$=0.67) but under predicted some of the major dust levels that were measured. The effect of temperature can be seen by the increase of both measured and modelled dust levels during

**Table 2** *Calculations of modelled dust levels vs measured dust levels for 30$^{th}$ May 2012 to 21$^{st}$ July 2012*

| Monitoring period end date | Days | Rain per day (mm) | Average temperature (°C) | RatioN (days of wind from N / days in MP) | Average northern wind speed (m s$^{-1}$) (speedN) | Wind risk (ratioN* speedN) | Modelled dust (EAC% / day) | Measured background dust (EAC% / day) |
|---|---|---|---|---|---|---|---|---|
| 30-May-12 | 7 | 0.58 | 22.6 | 0.14 | 4.5 | 0.64 | **1.51** | *0.48* |
| 06-Jun-12 | 7 | 0.04 | 24.1 | 0.29 | 9.6 | 2.76 | **3.16** | *1.87* |
| 13-Jun-12 | 7 | 0.22 | 26.5 | 0.43 | 14.6 | 6.26 | **5.67** | *5.30* |
| 22-Jun-12 | 9 | 4.26 | 26.0 | 0.11 | 11.9 | 1.32 | **1.56** | *1.61* |
| 27-Jun-12 | 5 | 0.00 | 25.6 | 0.20 | 3.5 | 0.69 | **1.94** | *0.22* |
| 04-Jul-12 | 7 | 0.44 | 23.6 | 0.43 | 8.2 | 3.51 | **3.54** | *3.71* |
| 07-Jul-12 | 3 | 0.00 | 25.9 | 0.33 | 10.7 | 3.57 | **3.87** | *8.32* |
| 10-Jul-12 | 3 | 0.00 | 28.4 | 0.67 | 8.4 | 5.62 | **5.46** | *6.56* |
| 14-Jul-12 | 4 | 0.00 | 27.0 | 0.25 | 3.9 | 0.98 | **2.25** | *0.51* |
| 17-Jul-12 | 3 | 2.29 | 27.4 | 0.67 | 7.8 | 5.18 | **4.62** | *5.08* |
| 21-Jul-12 | 4 | 0.13 | 27.8 | 0.75 | 9.4 | 7.02 | **6.29** | *6.97* |
| 24-Jul-12 | 3 | 0.00 | 28.2 | 0.00 | | 0.00 | **1.72** | *0.01* |
| 31-Jul-12 | 7 | 0.00 | 28.5 | 0.29 | 9.0 | 2.58 | **3.45** | *3.43* |

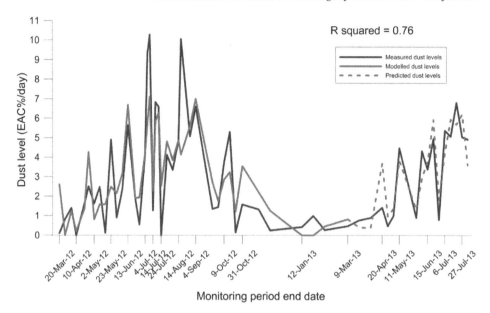

**Figure 6** *Measured dust levels plotted against modelled and predicted dust levels for the full 18 month monitoring period. Predicted dust levels are shown with a dotted red line.*

the summer period and subsequent drop from Oct-12 onwards. Wind speeds and proportions from the north cause dust levels to change erratically over short spaces of time in the summer and this proved difficult to fully capture in the model. Rainfall had little impact on predictions other than supressing modelled dust levels during heavy periods of precipitation.

The model was evaluated against recorded dust levels over the next six months. Weather data from the following six month period was input into the regression model and a prediction was made for dust levels. Background dust levels were then compared with positive results (Figure 6). The dust levels predicted showed high levels of accuracy ($r^2$=0.76), presenting with a better fit to the data than the original. Each peak and trough was predicted although some were underestimated and some were overestimated.

## 5.4 Summary of modelling study

Overall the modelling study has shown significant potential for the prediction of dust movements at and around industrial sites from a basic meteorological data set. One main limitation of the method appears to be in the modelling of the highest values; this may be attributed to the limitations of linear regression modelling. Linear regression modelling can by definition only model values with a linear relationship, and wind speed may have a cubic relationship with dust movements. This is because wind energy is proportional to the cube of wind speed[9]; small changes in wind speed can result in a big change in wind energy. The method could therefore be adapted to include a non-linear coefficient for wind speed in order to predict dust during periods of high wind speeds more accurately. Future developments can also include incorporating full scale site engineering and construction operations into the model. There also appears to be the potential for developing a general procedure for bespoke modelling of dust dispersion from other industrial, construction and dust generating sites. Modelling could be based on routinely available, meteorological data in addition to samples collected from dust monitoring as part of base-line environmental investigations.

## 6 CONCLUSIONS

This study has shown that by combining dust characterisation and dust modelling it is possible to understand both the content of collected dusts (and thus their origin), and the natural causes for their movements. The characterisation study revealed that dusts and local soils were essentially the same, whilst the modelling trial indicated that most dust movements from the north could be attributed to specific weather conditions. It may therefore be possible to predict dust movements at the site in addition to knowing the mineralogical and elemental make-up of the dust.

### Acknowledgements

We are grateful to Professor Geoffrey Walton for his support and encouragement of this work and to the Technology Strategy Board for funding the Knowledge Transfer Partnership for which made a proportion of this work possible.

### References

1  BSI, *BS 6069 Part 2. Characterisation of air quality, Part 2. Glossary,* British Standards Institution, 1994
2  Department for Environment, Food and Rural Affairs, *The Air Quality Strategy for England, Scotland, Wales and Northern Ireland,* HMSO, London, 2007.
3  Institute of Air Quality Management (IAQM), *Guidance on Air Quality Monitoring in the Vicinity of Demolition and Construction Sites,* Institute of Environmental Sciences, London, 2012
4  Department for Communities and Local government (DCLG), *National Planning Policy Framework,* HMSO, London, 2012
5  Minerals Industry Research Organisation, *Good practice guide: control and measurement of nuisance dust and PM10 from the extractive industries,* AEA Technology, Oxford, 2011
6  AEA UK Emissions Inventory Team (2008), UK Emissions of Air Pollutants 1970 to 2006, AEA Technology plc, Didcot.
7  H. Datson, M. Fowler and B. Williams, *Environmental Forensics,* Royal Society of Chemistry, Special Publication No. 338. Eds. Morrison, R.D and O'Sullivan, G., 2012.
8  M. Fowler, H. Datson, B. Williams and J. Bruce in *Environmental Forensics,* Royal Society of Chemistry, Special Publication No. 348. Eds. Morrison, R.D and O'Sullivan, G., 2013.
9  J. Andrews and N. Jelley, *Energy Science: Principles, Technologies, and Impacts,* Oxford University Press, UK, 2013.

ASSESSMENT OF POLYCYCLIC AROMATIC HYDROCARBONS IN AN URBAN SOIL DATASET

Rory Doherty, R. McIlwaine, L. McAnallen and S. Cox

School of Planning Architecture and Civil Engineering, Queen's University Belfast, UK BT9 5AG

1 INTRODUCTION

Polycyclic Aromatic Hydrocarbons (PAHs) are a group of semi-volatile organic compounds (SVOCs) that are composed of two or more aromatic (benzene) rings fused together in a variety of configurations. PAHs are widely distributed in the atmosphere and in urban soils and are one of the first pollutants to have been identified as suspected carcinogens[1]. In general, as the molecular weight of PAHs increases so does carcinogenicity, however acute toxicity, solubility and mobility decreases[2].
PAHs found in the urban environment are predominantly anthropogenic in origin. They may be either associated with (a) petroleum products introduced to the environment through spills and industrial discharges (generally termed "petrogenic") or (b) the incomplete combustion of organic materials emitted by various engine types from automobiles to power plants (termed "pyrogenic"). Low molecular weight alkyl PAHs generally indicate petrogenic origin (alkylated PAHs), whereas high molecular weight generally indicates pyrogenic origin (parent PAHs). The ratio of similar mass PAH isomers can give an indication of their likely source. PAHs from high temperature combustion sources are less thermodynamically stable than those from lower temperature sources[3].
Fugacity modelling can help to ascertain whether the ratios of the pyrogenic and petrogenic dominated PAHs may have altered in the soil phase over time or have been lost due to volatilisation during sampling. Fugacity is the tendency of a compound to "flee" or partition from one phase or compartment to another, and is measured in units of partial pressures (Pa). Environmental compartments (e.g. soil, gas, water) that are in equilibrium will have equal fugacity values. Each compartment will also have a fugacity capacity Z that takes into account concentration, temperature and physiochemical properties. Z is measured in $mol/m^3Pa$ and expresses the affinity of a chemical for a particular compartment[4]. This multimedia mass balance approach makes fugacity modelling an attractive method to provide insights into the environmental fate and transport of pollutants. Here we consider a suite of PAHs in soil that have been sampled across an urban setting. They are compared with airborne sampling and fugacity modelling in order to determine their likely source and equilibrium within the soil.

Source apportionment and fugacity modelling are vital first steps to allow the calculation of 'typical' concentrations of PAHs in soil by identifying the controlling factor/s over their spatial distribution. Once these factors are recognised, methods for calculating background (or baseline) concentrations of PAHs in soil can be applied e.g. the UK Normal Background Concentration[5] (NBC) and the Finnish Upper Limit of Geochemical Baseline[6] (ULBL). These approaches have differing philosophies behind the methods employed to capture background or baseline concentrations in line with their differing legislative frameworks. The UK approach tries to encompass as much anthropogenic input as possible into the normal background concentration where the Finnish ULBL approach is more conservative[7].

The objective of this research is to identify the main controlling factors over PAHs in an urban environment by applying source apportionment techniques such as fugacity modelling and isomeric ratios. Consideration will be given to the expected concentration of the PAHs within the soil, in order to assess how this urban setting would sit within a contaminated land legislative framework.

## 2 ASSESSMENT METHODS AND RESULTS

### 2.1 Data Sets

Two data sets were used for the evaluation reported here. One provided data on PAH concentrations in urban soil samples, and the other on air monitoring data. As part of a regional geochemical and geophysical survey, the TELLUS project, (http://www.bgs.ac.uk/gsni/tellus/overview/), undertaken by the Geological Survey of Northern Ireland, collected and analysed 30,000 soil, stream sediment, and stream water samples between 2004 and 2006. The TELLUS project is funded by the Department of Enterprise Trade and Investment for Northern Ireland and supports the exploration for, and development of, mineral and hydrocarbon resources, informs land-use planning and provides a country-wide environmental baseline. The urban soil data set used here provides an environmental baseline that may be of use to policymakers and regulators who are interested in land quality issues. Data relating to air monitoring of PAHs in the Belfast Metropolitan area is freely available from the UK Defra site (http://uk-air.defra.gov.uk/data/pah-data).

*2.1.1 Soil Sample Collection and analysis.* Regional or rural soil samples for inorganic analysis were collected at a density of 1 per 2 $km^2$ and 4 per $km^2$ for urban soil samples. As part of the urban sampling program additional samples for organic compounds were also collected at a density of 1 per $km^2$ and analysed for semi-volatile compounds including PAHs by GC-MS (Alcontrol laboratories, Chester UK). The urban sampling methodology followed the GBASE protocol where a shallow soil sample was taken from 5-20cm depth and composited from five locations at the corners and centre of a 20m x 20m grid with decontamination procedures adhered to at every location and samples stored at 4°C [8]. Obvious point sources of contamination were avoided and duplicate sampling was undertaken at 1 in 5 sample locations[9]. The duplicate sample grid was constructed adjacent to the original grid and analytical replicates were subsequently subsampled from the duplicate sample[10]. The Belfast Metropolitan Area (363 sample locations) and Londonderry (66 sample locations) were sampled as part of the urban organic campaign. A total of 792 samples were analysed with 61 locations used for replicate and duplicate analysis.

Airborne PAH data was selected from the only available Andersen Sampler located in Belfast from a similar time frame (2001- 2003). Detailed discussion of the air monitoring programme and analysis of PAHs in the Belfast area has been reported elsewhere[11,12].

*2.1.2 Soil Data Quality.* For the soil sample analysis, the laboratory (Alcontrol Geochem) reported the majority of the PAHs within the SVOC suite were within precision (15%) and bias (30%) required for UK MCERTs validation[13] with the exception of benzo[*a*]anthracene and naphthalene which were within the laboratory's own standards of 20% precision and 40% bias[14]. This was determined primarily using spiked samples alongside Certified Reference Material (CRM) testing of which only 2 of the 4 CRMs *117-100, 124-100* from RTC (www.rt-corp.com) performed well. This was attributed to milling of the CRM which may not reflect routine samples submitted resulting in a poor matrix match with the soil media[15]. Overall the values fell within the confidence intervals and prediction intervals given for CRMs such as *114-100* (Semi-VOAs in soil[16]). Further discussion of uncertainty around sampling and analysis is in the next section.

## 2.2 Apportionment of Uncertainty within Soil Samples

To attribute potential sources of error to either sampling, sub sampling and analysis, or geochemical factors a robust ANOVA[17] program RANOVA[18] was applied to duplicate-duplicate and duplicate-replicate pairs. This allows estimation of contribution of uncertainty from field sampling (from field duplicates) and lab sub-sampling and analysis (from field duplicates and laboratory replicates). Of the 61 sample locations chosen for duplicate and replicate analysis only 11 locations had a full group of 2 duplicates and 2 replicates with levels of at least one PAH that were above the detection limit of 100ug/kg. The robust ANOVA output is presented in Table 1 and it suggests that error can be primarily attributed to field duplicate sampling, followed by lab splitting and analysis, then geochemical variance. The generally low geochemical variance of four ring and greater PAHs can be explained by the source apportionment which identifies that they are predominately from a diffuse airborne combustion source rather than point source hotspots, and fugacity modelling (Section 2.3 - 2.4) that suggests the heavier PAHs are all relatively stable in soils. The fugacity modelling and lesser relative proportions of the lighter 2 and 3 ring PAHs suggest that these compounds are prone to sampling and laboratory preparation losses. A typical duplicate–replicate comparison such as 571327 and 571348 yielded values which varied from the mean by 29% acenaphthene, 16% anthracene, 16% benzo[*a*]anthracene, 10% benzo[*a*]pyrene, 10% benzo[*b*]flouranthene, 6% benzo[*g,h,i*]perylene. Here we see the percentage difference from the mean decreasing as molecular weight increases which may signal the loss of more volatile compounds through sample handling and preparation[19].

Thompson Howarth Plots also provide insight into attribution of uncertainty. Comparison of Thompson Howarth Plots for duplicate – duplicate (field based sampling) compared to duplicate replicate (lab splitting and analysis) suggesting that precision for both of these approaches was poor. Further comparison of duplicate and replicate samples of all PAHs on a summary Thompson Howarth plot where relative percentage difference from the mean was plotted against the mean also highlighted trends where sample pairs could be identified by the clustering of PAHs as they increased in molecular weight (Figure 1). Where such trends could be identified the maximum and minimum relative percentage differences of the stable PAHs in soils as they increased in molecular weight gave an indication of an analysis error within the overall sub sampling error. This

Assessment of Polycyclic Aromatic Hydrocarbons in an Urban Soil Dataset    95

**Table 1.** *Robust ANOVA of duplicate and replicate samples highlighting that the majority of the variance comes from a poor duplicate sampling approach*

|  | Number of polycyclic rings | Percentage of total variance | | |
|---|---|---|---|---|
|  |  | Across Sites | Field Sampling | Lab splitting and analysis |
| Phenanthrene | 3 | 2.20 | 91.52 | 6.28 |
| Fluoranthene | 4 | 21.81 | 67.62 | 10.57 |
| Pyrene | 4 | 11.05 | 73.94 | 15.01 |
| Chrysene | 4 | 0.00 | 83.45 | 16.55 |
| Benzo[*a*]anthracene | 4 | 0.00 | 84.61 | 15.39 |
| Benzo[*b*]flouranthene | 5 | 8.39 | 77.10 | 14.51 |
| Benzo[*k*]flouranthene | 5 | 27.81 | 57.94 | 14.26 |
| Benzo[*a*]pyrene | 5 | 2.42 | 86.45 | 11.12 |
| Benzo[*g,h,i*]perylene | 6 | 1.32 | 88.06 | 10.63 |
| Indeno[1,2,3-*cd*]pyrene | 6 | 0.00 | 58.52 | 41.48 |

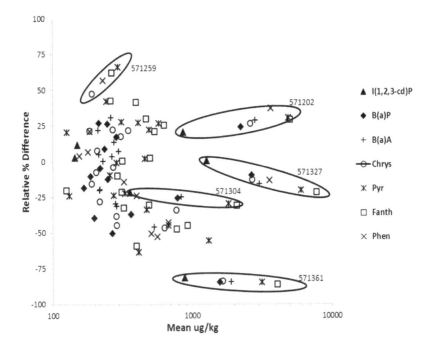

**Figure 1** *A modified Thompson and Howarth plot showing relative percentage difference against the duplicate-replicate mean for comparison. Some duplicate replicate pairs where easily identifiable are circled, e.g. 571361 refers to the pair of 571361 and 571383*

suggested that both duplicate sampling in the field and replicate sub sampling in the laboratory both could significantly contribute to errors. Comparison of one duplicate-replicate pair 571361, 571383 revealed concentration differences that ranged from 80 to 86 % of the mean across individual PAH compounds. Of this, the maximum to minimum range of relative percentage differences for the suite of individual PAHs analysed in the sample pair was only 6 %. This suggests that this error was due to lab splitting and creation of the replicate which provided a sub lot of very different concentrations. This suggests that neither the methodology for sampling duplicates in the field nor that for sub sampling replicates in the lab are robust and this method shouldn't be used for quality assurance testing of organic materials in urban areas until a validated sampling and sample preparation protocol is developed. Additional homogenisation in the field would lead to losses of the lighter volatile compounds as suggested by fugacity modelling (Section 2.4) but would have reduced sampling variance with the more stable PAHs. Other methods such as a field measurement duplicate approach [20] to reduce uncertainty or specialist sampling approaches [21-23] may need to be considered

### 2.3 PAH Characteristic Ratios and Cross Plots

To assist in determining the source of PAHs current practice is to compare the ratio of the less stable pyrogenic dominated isomer with its more stable petrogenic counterpart. This is often expressed in the form $C_x/(C_x+C_y)$ with the numerator acting as the less stable combustion dominated isomer[24]. The ratios are plotted as cross plots or bivariance plots to provide a more robust estimation of sources[25]. For four ring and larger PAHs boundaries are often drawn to differentiate between a fossil fuel, liquid fuel combustion and biomass / solid fuel combustion. The cross plots of ratios in Figure 2 compare soil and air monitoring data for selected PAH isomers[26] in the Belfast metropolitan area. Figure 2a shows the majority of soil samples suggest a combustion derived source in agreement with the air monitoring samples. This suggests that the ratios used in Figure 2a for soil samples have not yet undergone significant aging through mechanisms such as photodegradation[27]. The flouranthene to flouranthene + pyrene Fl/(Fl + Py) ratio suggests that the soil samples are dominated by the combustion of biomass and solid fuel whereas the indeno[1,2,3-cd]pyrene to indeno[1,2,3-cd]pyrene + benzo[ghi]perylene IP/(IP + BghiP) ratio suggests that the source is derived from fossil fuel combustion. However, it has been noted by other authors that it is difficult to differentiate liquid fuel and solid biomass combustion using this ratio[1]. Figure 2b uses the benzo[a]anthracene to benzo[a]anthracene + chrysene BaA/(BaA+Ch) ratio for comparison with the Fl/(Fl + Py) ratio, but in this case the air and soil samples do not plot within the same region with a mixed source and fossil fuel signature suggested for the air monitoring samples. The airborne monitoring samples are a 'fresher' source of contamination that has not yet been affected by depositional processes in soils. We also need to consider the similarity in thermodynamic stability of the BaA and Ch isomers of mass 228 which may limit their applicability as source ratio indicators[3] as well as the extraction efficiency which may co elute triphenylene alongside chrysene.

### 2.4 Fugacity Modelling

Due to the inability of the cross plots to clearly define particular sources potentially due to sampling errors that may cause the loss of volatiles by composite sampling, fugacity modelling was considered. In this case we pose a simple question about equilibrium of the PAHs in our soil samples across an urban area. Are the PAHs in soils likely to be affected

Assessment of Polycyclic Aromatic Hydrocarbons in an Urban Soil Dataset 97

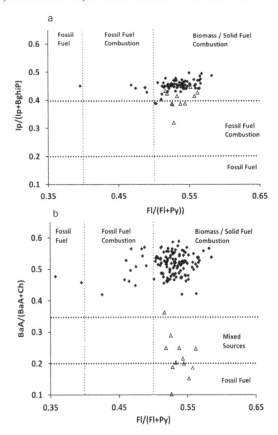

**Figure 2** *Cross Plots of characteristic ratios of PAHs that compare TELLUS urban soil samples (black diamonds) with quarterly urban air samples (triangles). The plots show proportions of a) Flouranthene to Flouranthene + Pyrene Fl/(Fl + Py) and indeno[1,2,3-cd]pyrene to indeno[1,2,3-cd]pyrene + benzo[ghi]perylene IP/(IP + BghiP) and b) Flouranthene to Flouranthene + Pyrene Fl/(Fl + Py) and Benzo[a]anthracene to Benzo[a]anthracene + Chrysene BaA/(BaA+Ch)*

by partitioning to other phases (soil gas, porewater) over time, causing the sampling approach to adversely affect the ratios that we use for source apportionment? The Equilibrium Criterion (EQC) model available from the Canadian Centre for Environmental Modelling (www.trentu.ca/cemc) was used. This model attempts to determine the fate of a chemical in a prescribed environment using a tiered approach that increases in complexity and input requirements. There are three tiers or levels, Level I (the approach used here) considers the equilibrium distribution or steady state in a closed environment and suggests the phases or environmental compartments that a chemical will partition to. Level I output identifies the compartment where the fugacity capacity Z and chemical concentrations are likely to be highest. Level II and III build on previous tiers by adding advective fluxes, degradation and transport between compartments, eventually allowing residence times to be calculated[28].

*2.4.1 Model Input Requirements.* A worst case scenario of 25°C or standard temperature and pressure (STP) was considered for the suite of PAHs that TELLUS urban sampling considered. It is highly unlikely that soil and groundwater temperature reflect these values but it could be used as a proxy to indicate a gross sampling error such as poor sample storage which may result in losses of more volatile PAHs from soils typical model input is highlighted in Table 2. The model output was benchmarked against published EQC output for benzo(a)pyrene (B[a]P) at 25°C[29] then rerun for the other PAHs. Temperature dependant input parameters were also calculated and rerun for the suite of PAHs at 16°C and 11°C which represented the maximum and mean typical soil temperatures for Northern Ireland[30].

*2.4.2 Model Output.* The model output (Figure 3a,b,c) shows the relative composition of PAHs in soil, water and air at 25°C, 16°C and 11°C. Four ring and larger PAHs from pyrene to indeno[1,2,3 –c,d]pyrene all partition strongly (>98%) to the soil compartment at 25°C, 16°C and 11°C. The source apportionment ratios used on this dataset {Fl/(Fl + Py), BaA/(BaA+Ch) and IP/(IP + BgP)} will not have experienced losses partitioning to other phases due to soil temperature effects. This stability of the heavier PAHs combined with the characteristic ratios suggests that we can say with confidence that the 4 ring and larger PAHs in soils are predominately from a diffuse combustion source. We cannot clearly define whether this source is from fossil fuel combustion (petroleum product combustion) or from solid fuel or biomass combustion. The Fl/(Fl + Py) and BaA/(BaA+Ch) ratios suggest that the soil samples are predominated by a solid fuel biomass combustion source, while IP/(IP + BgP) suggests a fossil fuel combustion source. This highlights the difficulty in clearly ascribing a generalised PAH ratio to all scenarios when often regional specific or site specific factors could have an effect. More lines of evidence need to be applied to specifically attribute the actual source(s) of combustion in the Belfast urban area.

**Table 2** *Example of typical EQC model input requirements for pyrene*

| Property | Value |
|---|---|
| Enthalpy of Vaporisation (J/mol) | 89400.00[31] |
| Enthalpy Of Fusion | 10200.00[31] |
| Molar Mass (g/mol) | 202.25[32] |
| Data Temp. °C | 25.00[30] |
| Melting Point °C | 150.25[32] |
| Vapour pressure (Pa) | 5.91E-04 [33] |
| Solubility in Water (g/m³) | 0.13[33] |
| Henry's Law Constant (Pa·m³/mol) | 0.92[33] |
| $\log K_{OW}$ | 5.18[32] |
| Air-Water ($K_{AW}$) | 1.60E-03[a] |
| Aerosol-Air | 5.86E+08[a] |
| $K_{OC \, (L/kg)}$ | 52974.64[34] |

[a] input calculations performed by the model

The model outputs at all temperatures also reflect the general distribution of PAHs in soils (Figure 2d). The soil PAH data is dominated by flouranthene and pyrene both in number of samples and in concentration. Lighter PAHs (naphthalene to flourene) are noticeably absent. This suggests that the lighter compounds are quickly lost from soils or were not present in any great quantities when they were deposited. When compared with air monitoring data from a similar period (Figure 3d) we see that the air data is dominated by phenanthrene followed by flourene then flouranthene and pyrene. This suggests that if PAHs are being deposited from an airborne diffuse source then the lighter compounds naphthalene to acenaphthalene are unlikely to be deposited with flourene and phenanthrene experiencing rapid losses as they are deposited[27]. Care must also be taken as these airborne figures represent only 3 years of quarterly data from one sampling location in the urban area and may not be representative of a regional urban distribution[11,12]. Both phenanthrene and flourene are compounds that our EQC model suggests will show some partitioning across a range of temperatures, phenanthrene is expected to have a 95-6% and Flourene 84-90% composition in soils so losses to air and water may be expected over time. From Figure 2 it can be seen that phenanthrene dominates the air samples but does not completely make the transition to soil samples. Based on the EQC partitioning and the differences between air monitoring and soils data care must be taken if considering the use of diagnostic ratios of PAHs of mass lighter than 178 (phenanthrene & anthracene) in the soil phase. There may also be significant differences between the air and soil phases as part of the deposition process.

## 2.5 Comparison of Calculation Methods to determine Anthropogenic Background Calculations

The urban soil PAHs from flouranthene to indeno[1,2,3-cd]pyrene are at equilibrium in the soil phase, they are not subject to partitioning losses and have characteristic ratios that are dominated by a combustion source. We can confidently assume the majority of the PAH soil samples in the urban area come from diffuse airborne source(s). Such a conceptual model of diffuse deposition to soils allows us to consider anthropogenic background or baseline concentrations of PAHs. Background concentrations of elements or compounds have often been applied to geogenic or 'natural' sources[35]. The advent of the term Anthropocene[36] alongside statutory approaches to land contamination[5] allows us to consider anthropogenic background or baseline levels of contaminants in the urban environment. In this case we can show that within the Belfast urban area airborne deposition of PAHs from combustion sources is the controlling factor. A considerable amount of individual PAHs in samples in the Belfast Metropolitan Area, as sampled using the GBASE methodology, are below the method detection limit of 100ug/kg. This makes the calculation of a background value for the whole urban area impossible because the calculation would be dominated by any value used as a replacement for the Method Detection Limit (MDL)[37]. We can depart from the standard Normal Background Concentration[5] (NBC) methodology for defining domains by using our previously defined characteristic ratios to create an urban combustion subdomain defined within the Belfast Metropolitan area. Here the characteristic ratios alongside fugacity modelling have defined an area where a single controlling factor can be attributed to the spatial distribution of a chemical where the samples show concentrations above the MDL. This calculation within an urban subdomain allows the screening out of areas below MDL and focuses attention onto areas of interest where sites above the calculated background values can be identified

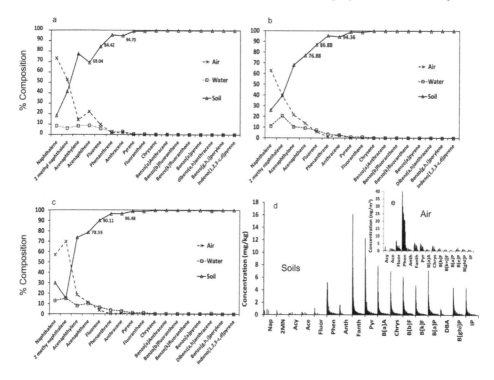

**Figure 3.** *EQC model output showing relative compositions of PAHs in Soil, Water and Air at a) $25°C$ b) $16°C$ c) $11°C$ and d) the results of urban soil & e) air sampling*

for future investigation. However care must be taken that the MDL is not so high that future decisions on land quality would be compromised. Statistical analysis was carried out using R software (http://www.r-project.org/). The UK NBC approach for calculating normal background concentrations was applied to our urban subdomain by first assessing the skewness of the data, secondly, performing a log or box-cox transformation of the data if necessary and thirdly calculating percentiles using parametric, robust or empirical approaches depending on the results of the previous transformation process. R scripts written by Cave *et al.* can be used to complete this process[38]. The Finnish approach to derive the ULBL normally defines the equivalent of a domain as a soil type within a geochemical province[6]. In this case within the urban subdomain we would attribute the soil type as 'made ground' and apply the same boundary or domain as the normal background concentration. The ULBL is the upper limit of the upper whisker of a box and whisker plot. Data which falls outside this are regarded as outliers that are probably not typical of the geochemical province. Unlike the UK NBC method the data is not transformed, and therefore the Finnish approach generates more outliers which are likely to be anthropogenic producing a more conservative background value. A comparison of the two methods is presented in Table 3. From Table 3 it is immediately obvious that the two methods provide grossly different anthropogenic background concentrations. This can be traced back to the underlying philosophies behind the methods controlled by their respective legislative regimes. The NBC concentrations are much higher than those calculated via the ULBL method (5.4, 7.5 and 3.6 times greater for B[*a*]P, Fl and Py respectively). The number of sample locations where concentrations are higher than the

**Table 3.** *Summary of values calculated by the ULBL and NBC methods for Benzo[a]pyrene B[a]P, Fluoranthene, Fl and Pyrene, Py*

|  |  | NBC calculation (µg/kg) | ULBL calculation (µg/kg) |
|---|---|---|---|
| B[a]P (n = 83) | Empirical percentiles | 5300 | 980 |
| Fl (n = 121) | Empirical percentiles | 9700 | 1300 |
| Py (n = 121) | Box-cox transform & theoretical percentiles | 4300 | 1200 |

NBC are 2 sites for B[a]P, 3 sites for Fl and 7 sites for Py. Using the ULBL method, this increases to 14,17 and 15 for B[a]P, Fl and Py respectively therefore providing a more conservative estimate of the number of sites that may require further investigation within a land quality assessment. When compared with other urban settings where NBCs have been calculated such as London[39] the values are similar (B[a]P 6mg/kg, Fl 9.7mg/kg, Py 8.4mg/kg) with B[a]P exceeding generic assessment criteria for residential land use (2.4 - 2.5mg/kg) and allotments (2.7mg/kg)[40] Care must be taken when using either of these values. A quality assurance assessment of the dataset suggests that while laboratory-reported precision and bias fall within an acceptable range for UK regulatory agencies, the methodology for field sampling of duplicate and laboratory splitting of the replicate samples didn't perform well. However, this dataset can be used to highlight locations that may require further investigation due to their elevated concentrations of certain PAHs.

## 3 CONCLUSIONS

The collection and analysis of large quantities of samples for semi-volatile compounds such as PAHs across urban environments is a difficult task. Sampling, compositing, sub sampling and analysis errors can make the data difficult to interpret[19]. The use of specialist sampling approaches to reduce uncertainty[20] or sampling equipment to ensure homogenisation and sample representativeness should be used in the field[41] and in the lab[23,42]. Tools like fugacity modelling allow estimation of how we can expect semi-volatile compounds to partition into or out of urban soils and what isomeric PAH ratios are likely to remain stable over time. Comparison of soil and air isomeric PAH ratios suggest that in this case urban soil PAHs are dominated by an airborne combustion source. The definition of an urban subdomain where combustion can be identified as the dominant controlling factor allows urban areas below MDL to be screened out and focuses attention onto background calculations in areas of interest. However care must be taken to ensure that the MDL is not so high that it clouds future decisions that could be made on land quality. The methodologies used here to calculate background concentrations across the same urban combustion subdomains vary considerably in their output. This may be attributed to differing philosophies of what is 'background', which is driven by legislative requirements. The fact that anthropogenic datasets such as semi volatile organic compounds are also typically highly skewed highlights difficulties of using methods that require transformation of data. The development of new analysis technologies such as GC-GC methods[43] and their uptake by industry will allow a more detailed suite of analysis that moves source apportionment analysis and interpretation into a new era. This combined with specially developed sampling approaches[21] including incremental sampling[22] for semi volatile organic compounds will allow detailed source apportionment and robust calculations of anthropogenic background or baseline concentrations across large urban areas.

**Acknowledgements**

The Northern Ireland TELLUS project was funded by the Northern Ireland Department of Enterprise, Trade and Investment and by the Rural Development Programme through the Northern Ireland Programme for Building Sustainable Prosperity. R.McIlwaine & L.McAnallen are funded by the Department of Education and Learning (Northern Ireland).

**References**

1. K. Ravindra, R. Sokhi, and R. Vangrieken, *Atmos. Environ.*, 2008, **42**, 2895–2921.
2. K.-H. Kim, S. A. Jahan, E. Kabir, and R. J. C. Brown, *Environ. Int.*, 2013, **60**, 71–80.
3. M. B. Yunker, R. W. Macdonald, R. Vingarzan, H. Mitchell, D. Goyette, and S. Sylvestre, *Org. Geochem.*, 2002, **33**, 489–515.
4. D. Mackay, J. A. Arnot, E. Webster, and L. Reid, *Ecotoxicol. Model.*, 2009, **2**, 355–375.
5. E. L. Ander, C. C. Johnson, M. R. Cave, B. Palumbo-Roe, C. P. Nathanail, and R. M. Lark, *Sci. Total Environ.*, 2013, **454-455**, 604–618.
6. J. Jarva, T. Tarvained, J. Reinikainen, and M. Eklund, *Sci. Total Environ.*, 2010, **408**, 4385–4395.
7. R. McIlwaine, S. F. Cox, R. Doherty, S. Palmer, U. Ofterdinger, and J. M. McKinley, *Environ. Geochem. Health*, 2014.
8. C. Johnson, 2005 G-BASE Field Procedures Manual. Internal Report, IR/05/097, Nottingham, UK, 2005.
9. K. Knights, A report on the Tellus urban field campaigns of Belfast Metropolitan areas and Londonderry, 2006. Commissioned Report CR/07/006N, 2006.
10. S. E. Nice, Inorganic soil geochemical baseline data for the urban area of the Belfast Metropolitan Area, Northern Ireland. Open Report OR/08/021, Nottingham, UK.
11. D. M. Butterfield and R. Brown, Polycyclic Aromatic Hydrocarbons in Northern Ireland NPL REPORT AS66, Middlesex, 2012.
12. A. S. Brown, R. J. C. Brown, P. J. Coleman, C. Conolly, A. J. Sweetman, K. C. Jones, D. M. Butterfield, D. Sarantaridis, B. J. Donovan, and I. Roberts, *Environ. Sci. Process. Impacts*, 2013, **15**, 1199–215.
13. HMSO, MCERTS: Performance Standard for Laboratories Undertaking Chemical Testing of Soil, 2014.
14. J. Jones, "Customer Query Investigation 06/10325/01," 2006.
15. Analytical Methods Committee, *Anal. Methods*, 2012, **4**, 3521.
16. GSNI, Investigation into Duplicate Analyses (11/10/06), 2006.
17. M. H. Ramsey, *J. Anal. At. Spectrom.*, 1998, **13**.
18. P. Rostron, *Anal. Methods*, 2014, **6**, 7110.
19. D. J. Beriro, C. H. Vane, M. R. Cave, and C. P. Nathanail, *Chemosphere*, 2014, **111**, 396–404.
20. M. H. Ramsey and K. A. Boon, *Geostand. Geoanalytical Res.*, 2010, **34**, 293–304.
21. B. Gustavsson, K. Luthbom, and A. Lagerkvist, *J. Hazard. Mater.*, 2006, **138**, 252–60.
22. P. W. Hadley and I. G. Petrisor, *Environ. Forensics*, 2013, **14**, 109–120.

23. R. . Gerlach, J. M. Nocerino, C. A. Ramsey, and B. C. Venner, *Anal. Chim. Acta*, 2003, **490**, 159–168.
24. M. B. Yunker, A. Perreault, and C. J. Lowe, *Org. Geochem.*, 2012, **46**, 12–37.
25. H. Chen, Y. Teng, and J. Wang, *Sci. Total Environ.*, 2012, **414**, 293–300.
26. U. M. Sofowote, L. M. Allan, and B. E. McCarry, *J. Environ. Monit.*, 2010, **12**, 417–24.
27. D. Kim, B. M. Kumfer, C. Anastasio, I. M. Kennedy, and T. M. Young, *Chemosphere*, 2009, **76**, 1075–81.
28. L. Hughes, D. Mackay, D. E. Powell, and J. Kim, *Chemosphere*, 2012, **87**, 118–24.
29. T. Ntakirutimana, J. Guo, X. Gao, and D. Gong, *Res. J. Environ. Earth Sci.*, 2012, **4**, 731–737.
30. A. M. García-Suárez and C. J. Butler, *Int. J. Climatol.*, 2006, **26**, 1075–1089.
31. M. V. Roux, M. Temprado, J. S. Chickos, and Y. Nagano, *J. Phys. Chem. Ref. Data*, 2008, **37**, 1855.
32. Y.-G. Ma, Y. D. Lei, H. Xiao, F. Wania, and W.-H. Wang, *J. Chem. Eng. Data*, 2010, **55**, 819–825.
33. N. Earl, C. Cartwright, S. Horrocks, M. Worboys, S. Swift, S. Kirton, A. Askan, H. Kelleher, and D. Nancarrow, *Review of the Fate and Transport of Selected Contaminants in the Soil Environment R&D Technical Report P5-079/TR1*, Bristol, 2003.
34. R. Seth, D. Mackay, and J. Muncke, *Environ. Sci. Technol.*, 1999, **33**, 2390–2394.
35. J. Matschullat, R. Ottenstein, and C. Reimann, *Environ. Geol.*, 2000, **39**.
36. J. Zalasiewicz, M. Williams, W. Steffen, and P. Crutzen, *Environ. Sci. Technol.*, 2010, **44**, 2228–31.
37. D. R. Helsel, *Chemosphere*, 2006, **65**, 2434–9.
38. M. R. Cave, C. Johnson, E. L. Ander, and B. Palumbo-Roe, Methodology for the determination of normal background contaminant concentrations in English soils Land Use Planning and Development Programme Commissioned Report CR/12/003, Nottingham, UK, 2012.
39. C. H. Vane, A. W. Kim, D. J. Beriro, M. R. Cave, K. Knights, V. Moss-Hayes, and P. C. Nathanail, *Appl. Geochemistry*, 2014, **51**, 303–314.
40. CL:AIRE, SP1010 – Development of Category 4 Screening Levels for Assessment of Land Affected by Contamination - Appendix E, 2014.
41. R. W. Gerlach, D. E. Dobb, G. A. Raab, and J. M. Nocerino, *J. Chemom.*, 2002, **16**, 321–328.
42. C. A. Ramsey and J. Suggs, *Environ. Test. Anal.*, 2001, **10**, 12–16.
43. G. Frysinger, R. B. Gaines, and C. M. Reddy, *Environ. Forensics*, 2002, **3**, 27–34.

STATISTICS OF LOST HISTORICAL COAL TAR PAH CONTAMINATION IN SOILS AND SEDIMENTS

Michael J. Wade

Wade Research, Inc., 110 Holly Road, Marshfield, MA 02050-1724, USA

1   INTRODUCTION

The purpose of this paper is to examine the use of polycyclic aromatic hydrocarbons (PAH) data in the investigation of coal gasification contaminants. In the United States legal system considerable demands are placed upon forensic geochemistry to unlock tangled sources of PAH at former coal gasification and coal tar processing sites. Time is not on the side of the forensic investigator. Things change. While there is a defined and well documented industrial time line, decades long, in Europe and in the United States, care must be exercised to not neglect the historical engineering record. Coal gasification operational impacts in the modern era arise from engineering process decisions that were made in the 1800s. Important engineering decisions were made to facilitate production of hydrocarbon gases for illumination and chemical manufacturing in the manufactured gas production (MGP) industry. Commensurate with coal gasification was coal tar management. In the United States, MGP development accelerated as the technology arrived from Europe in the middle of the nineteenth century. Changes in gasification technology over time carried operational manifestations that also carried long-term consequences that were not recognized nor appreciated at the time. Changes in gasification technology from batch operations to technologies such as horizontal coal gas retorts, carbureted water gas, inclined gas retorts, coke oven installations and oil gas each carried attendant hidden contaminant consequences [1-3] that were, perhaps at the time, viewed as unimportant.

Increased availability of natural gas in the United States in the early twentieth century, driven by improved petroleum exploration and better pipeline technology developments, made use of coal for the MGP industry unattractive. Eventually, MGP businesses passed out of use and were ended. However, the changing energy technologies did nothing to diminish the value of the properties where former MGP operations were conducted. As MGP operations almost by necessity were proximal to their customers, i.e., developing major population centers, the associated properties maintained or increased in value. Accordingly, properties were redeveloped, repurposed in a variety of ways and their use continued. In the mid-twentieth century, most repurposing remained industrial with the properties owned by MGP successor companies and were associated with redevelopment for better electrical generation and/or petroleum operations, etc. Redevelopment of such properties changed again in the later part of the twentieth century as "heavy industry" properties were once again redeveloped for different commercial uses and, in response to increased population growth and their proximal location to major population centers, into residential properties.

Throughout the life of the MGP industry there were changes in technology that had commensurate consequences in both MGP products as well as wastes. MGP products were sold and wastes were discarded. Eventually, when uses were developed for chemicals in waste materials, they were sent for reprocessing into increasingly more valuable materials. Moreover, it is fair to say that remediation of residual MGP wastes was not the top priority during property redevelopment in the early to mid-twentieth century. As other industrial development occurred, a different mix of wastes was generated and either dealt with or not. Therefore, for each change in property use there was a change in the mix of attendant contaminants. Over time, various wastes became mixed at these properties. Eventually as U.S. environmental laws and regulations were put in place in the 1970s and early 1980s, as well as elsewhere, waste issues had to be dealt with in a more formal manner. Now, in the United States, formal debate of property contamination liability issues often occurs in the courtroom. PAHs are prominent but by no means sole contaminants of environmental courtroom for MGP site assessments.

PAHs derived from coal, coal gasification and coal tar wastes were rarely dealt with at MGP sites. Such was especially true during the early development of the MGP facilities because wastes were perceived as engineering problems that interfered with gas production goals and were dealt with by modifying engineering processes to continue or increase gas production. As property redevelopment progressed in the twentieth century, resident wastes termed legacy MGP wastes were overridden by a more modern mix of wastes such as those derived from the use of petroleum-derived fuels at these sites. Prominent among the petroleum derived wastes were PAHs. Sorting out whose PAHs are whose is one of the major issues with which the legal proceedings deal, hence the need for a forensic source assessment to PAH geochemistry at MGP sites.

There has been some forensic effort put into discerning which PAH contaminants are indicative of which coal gasification technologies in an effort to aid in the assignment of financial liability for environmental remediation cleanup. This statement is especially true for carbureted water gas versus gas oil wastes.[4] But there were other MGP technologies that may have had an effect on the generation of PAH wastes too. Therefore, in any single investigation, it is important to match the waste composition with the record of site technologies used at the property. Unfortunately, matching industry wide MGP technology changes to site specific investigations often require site documentation that may have long since cease to exist. In the absence of such information, assumptions are sometimes made. However, while the making of unsupported assumptions about the industry wide availability of selected MGP technologies as they apply to a specific site is not recommended, such assumptions are made nonetheless.

It is clear that analytical laboratory tools available to the forensic geochemist can keep pace with the legal system's demands, but who leads whom is sometimes less clear. Gas chromatography (GC), either combined with flame ionization detection (GC/FID) or combined with mass spectrometry (GC/MS), is the tool that most often is applied for investigation of PAH contamination from coal gasification/coal tar processing operations. Various PAH analyte lists exist. Analytical approaches include lists of the basic U.S. Environmental Protection Agency (EPA) priority pollutant PAHs, as well as more comprehensive parent PAHs and their alkylated homologues. Table 1 provides a list of each PAH category. There is one subset of the list that is comprised of established EPA priority pollutant PAHs, depending, of course, on whether or not methylnaphthalene is included on the analyte list. However, there is no standard analyte list of parent and alkyl homologue PAHs, as each laboratory - and it seems each investigator - has their own subset PAH list. Boehm [5] discusses the variability of the parameter total PAH ($\Sigma$ PAHn) where n is the number of individual PAHs quantified. A range of total PAH ($\Sigma$ PAHn) values are discussed. The fact remains that variable $\Sigma$ PAHn approaches, driven by analytical laboratory offerings, should not be considered settled science. Further, it should not be assumed that $\Sigma$ PAHn is an

**Table 1** *List of U.S. EPA Priority Pollutant PAHs and A Second Separate List of Parent and Alkylated PAHs from a U.S. Forensics Analytical Laboratory*

| U.S. EPA Priority Pollutant PAHs | | | |
|---|---|---|---|
| Acenaphthene | Acenaphthylene | Anthracene | Benz[a]anthracene |
| Benzo[a]pyrene | Benzo[e]pyrene | Benzo[b]fluoranthene | Benzo[ghi]perylene |
| Benzo[j]fluoranthene | Benzo[k]fluoranthene | Chrysene | Dibenz[a,h]anthracene |
| Fluoranthene | Fluorene | Indeno[1,2,3-cd]pyrene | Naphthalene |
| Phenanthrene | Pyrene | | |
| **Alkylated PAH** | | | |
| Decalin | C1-Decalins | C2-Decalins | C3-Decalins |
| C4-Decalins | Naphthalene | C1-Naphthalenes | C2-Naphthalenes |
| C3-Naphthalenes | C4-Naphthalenes | 2-Methylnaphthalene | 1-Methylnaphthalene |
| Benzothiophene | Biphenyl | 2,6-Dimethylnaphth | Dibenzofuran |
| Acenaphthylene | Acenaphthene | 2,3,5-Trimethylnaph | Fluorene |
| C1-Fluorenes | C2-Fluorenes | C3-Fluorenes | Dibenzothiophene |
| C1-Dibenzothiophenes | C2-Dibenzothioph | C3-Dibenzothioph | C4-Dibenzothioph |
| Anthracene (Anth) | 1-Methphenanthrene | Phenanthrene (Phen) | C1-Phen/Anth |
| C2-Phen/Anth | C3-Phen/Anth | C4-Phen/Anth | Fluoranthene (Fluoranth) |
| Pyrene | C1-Fluoranth/Pyrenes | C2-Fluoranth/Pyrenes | C3-Fluoranth/Pyrenes |
| C4-Fluoranth/Pyrenes | Naphthobenzothio | C1-Naphthobenzothio | C2-Naphthobenzothio |
| C3-Naphthobenzothio | Benz[a]anthracene | Chrysene | C1-Chrysenes |
| C2-Chrysenes | C3-Chrysenes | C4-Chrysenes | Benzo[b]fluoranthene |
| Benzo[k]fluoranthene | Benzo[e]pyrene | Perylene | Benzo[a]pyrene |
| Indeno[1,2,3-cd]pyrene | Dibenz[a,h]anth | Benzo[g,h,i]perylene | |
| 17a(H),21B(H)-hopane - C30H52 | Carbazole | | |
| NOTE: Dimethylnaphth = Dimethylnaphthalene; Trimethylnaph = Trimethylnaphthalenes; Dibenzothioph = Dibenzothiophenes; Phen = Phenanthrene ; Anth = Anthracene; Naphthobenzothio = Naphthobenzothiophene | | | |

independent variable in and of itself because for one set of samples a single individual PAH could be dominant, while for another set of samples a different PAH or set of PAHs could be dominant. For the investigator believing that more data is better, the larger the n value, the better the dataset.

Suffice it to say that, if desired and for that matter if paid for, today's modern analytical technologies can deliver a considerable array of information, on a per sample basis, concerning MGP investigations. However, there is a growing trend in the forensic geochemical community to focus down both the field and laboratory efforts, obviously with attendant reductions in expense, and to reduce data analysis to indicator ratios that are purported to tell an investigator all that there is to know about MGP PAH wastes and attendant environmental contamination quickly and cheaply. Accordingly, in light of these recent trends, it seems that an examination of laboratory data generation and attendant data analyses approaches is in order as it concerns MGP and PAHs.

## 2    SOURCES OF PAH IN THE ENVIRONMENT

Blumer [6] dealt with sources of PAHs in the natural environment and stated that high temperature processes should result in PAH assemblages that would be devoid of alkyl PAHs. Neff [5] in discussing the formation of PAH by pyrolysis of organic matter, reached a somewhat different conclusion regarding coal tar processes and the presence of PAH alkyl homologues in PAH assemblages. In talking about coal tar production, Neff [7] stated:

" . . . gaseous emissions are enriched in lower molecular weight PAH in comparison to coal tar, while coal tar has relatively higher concentrations of higher molecular with PAH from the emissions. Alkyl PAH are well represented in both coal tar and the emissions but their concentrations are generally much lower than those of the unalkylated parent PAH."

One obvious issue to note in Neff's conclusions is that much lower does not equal absent. Also, it should be noted that significant advances in gas chromatography since the early to mid-1970s have been made in both reliability of analyte identification as well as an increase in sensitivity of detection that are matrix independent, so, perhaps, the positions of Blumer and Neff were not in such an apparent conflict. As emphasized by Boehm, [5] temperature plays a significant role in the presence/absence of alkylated PAHs and without understanding specific engineering conditions of a MGP facility, making generalizations about what might or might not be present from MGP operations not scientific. Indeed, as data cited by Boehm [5] from Neff et al.[8] illustrate, alkyl PAHs are present in MGP waste and byproducts. Therefore, it is important to emphasize that simply assuming MGP sites and coal tar wastes had no alkyl PAH assemblages at all without generating sustaining laboratory data is assumptive and inherently dangerous. Moreover, as was often the case, over time petroleum-based contamination became mixed with MGP wastes. It is accepted that petroleum products contain parent as well as significant amounts of alkyl PAHs.[5,7] Therefore, use of laboratory data on the parent and alkyl homologue PAH analytes to separate out petrogenic and pyrogenic process signatures on PAH assemblages is warranted.

It is becoming increasingly more common in MGP investigations that the analytical effort is directed only to parent PAHs and not to alkyl PAHs. Analytical laboratory cost savings result. Moreover, an assumptive data analysis approach to focus only on the higher molecular weight PAHs as being indicative of coal tar from MGP operations seems to appear from time to time. Under such circumstances, a common rationale for focusing only on higher molecular weight PAHs is to eliminate confounding influences of environmental weathering processes (i.e., the combined processes of evaporation, solubilization, adsorption, microbial degradation, etc.) by ignoring the lower molecular weight PAHs in the analyte dataset. Unfortunately, in doing so, it appears that valuable information on MGP process impacts and coal tar management approaches at specific sites may be lost.

## 3 MGP IMPACTS – THE ROLE OF NAPHTHALENE

Of particular concern in the coal gasification and coal tar industries was the occurrence of naphthalene. Historically, naphthalene was viewed as a problem.[9-13] Naphthalene was an unintended manufactured byproduct of the gas production process during the retort of coal. Coal gases for illumination, etc. were the intended end product. Unintended gasification products included water vapor, sulfides, cyanides, and ammonia, but waste proportions varied from process to process.[1] Naphthalene was also an unintended gasification product as well. Moreover, from an engineering standpoint, naphthalene was to be avoided due to its particular thermodynamic property of existing either as a gas or as a solid at common process temperatures and pressures. Indeed, it is not until pressures of above 60 atmospheres are created at +25 °C that naphthalene remains in the liquid state. It is easy to see, therefore, that if coal gasification conditions were not optimal, the naphthalene created by chemical condensation reactions within the gas phase from smaller molecules would drop out of the gaseous phase directly into a solid phase and create a blockage in the system. Such behavior resulted in a "naphthalene plug" and was often referred to in the technical literature of the nineteenth century as the naphthaline (naphthalene) problem; perhaps a most graphic description in referring to naphthalene was a statement as to . . . "thorn in the flesh" . . . of the gas manufacturing industry.[14] Moreover, seasonal problems from low winter temperatures associated with naphthalene plugged pipes were also documented for some sites,[15] but not for others.[16]

Thus, early in the life of the coal gasification industry, creation of naphthalene was to be avoided or when produced, naphthalene was to be kept in the gaseous phase or discarded in just about any possible manner. At some sites, naphthalene scrubbers became common and naphthalene was discarded with the product water and/or coal tars. Such products initially were simply discarded in the vicinity of the manufacturing site. It was not until later in the nineteenth and earlier in the twentieth century that the utility of such byproducts began to be appreciated and reclamation processes were designed and implemented. At other sites, naphthalene was not a problem due to maintaining of careful gas condensation conditions.

What does the presence of naphthalene in MGP wastes tell us about coal gasification processes at a site? To simply ignore the presence of naphthalene as unrelated to the MGP process clearly is wrong. But is naphthalene a sole indicator chemical for coal gas production or coal gas waste streams or, for that matter, petroleum? The short answer is no to all. Naphthalene can serve as an indicator of MGP processes and methods (or lack thereof) of handling MGP wastes (e.g., coal tars and scrubber wastes). At best, it appears that naphthalene may serve as both an indicator chemical for gas manufacturing as well as a chemical associated with waste management practices at a MGP site at different times in the life of a facility. Serving in either role, ignoring naphthalene in PAH data does not assist in the documentation of the environmental consequences of coal gasification processes, whatever the basic engineering manufacturing processes may have been. In such a dual role, it may be tempting to ignore naphthalene in the PAH analyses data because its presence in the dataset complicates the understanding of the data. But ignoring a part of the PAH dataset is not defensible either. Simply assuming that naphthalene is present solely due to petroleum product influences is similarly wrong. However, as can be seen in the case studies in the following section, the answers obtained by including naphthalene in the PAH data vary. As might be expected, there is no one simple answer derived from an examination of the "naphthalene issue" at former MGP properties.

## 4   SOURCE ANALYSIS OF PAH DATA

Published technical literature on source allocation offers numerous ways of using PAH data (e.g., see references [4,5,17,18]). The cited published scientific works are comprehensive, employing multiple ways of analyzing PAH data. Double ratio plots using both alkylated PAH data combined with parent PAH data used provide a more comprehensive approach to data analyses than single ratio plots, for example. Limiting the PAH analytes to parent PAHs restricts the data analyses to parent PAH data plots, thereby losing some amount of information. Recent experience has demonstrated that single ratio plots are being used to sort out MGP influences more freely than one should expect. The net effect of using only a plot of 4- and 5-ring PAHs in analyzing data eliminates information that can prove informative to a MGP site investigation as illustrated in the following three case studies.

### 4.1   Source Analysis – Case Study Number 1

For Case Study Number 1, significant environmental impacts from operations at a MGP plant and associated coal tar reprocessing facility could be evaluated from data collected as a consequence of industrial property redevelopment to residential uses in a major US metropolitan city. The site took in coal tar wastes from a variety of MGP facilities and engaged in the manufacture of roofing materials, recovery of chemicals serving as raw materials for further processing, benzole, naptha, naphthaline and coal-oil as well as other fuel materials. In this case, documentation of PAH impacts during site remediation were centered on regulatory compliance (e.g., public health driven) related issues and assignment of source(s) came as an after-the-fact issue. Before forensic source assessment was considered, all site work and analytical work had been completed and numerous laboratory data packages were delivered to the contracting entities. Available site documentation included information regarding products manufactured, layout diagrams for various engineering processes, and locations of incoming material and product storage tanks. Throughout the investigation over 400 groundwater, soil and sediment samples had been collected and analyzed for U.S. EPA priority pollutants while less than five soil samples were collected and analyzed for parent and alkyl PAHs. PAH data came from multiple analytical laboratories. After all the field and lab reports were issued, almost as an afterthought, forensic analysis was sought to support legal action to recover costs and damages incurred and will incur to investigate and remediate historic environmental contamination. Redress for damage caused by MGP coal tar reprocessing at the property was sought under the U.S. EPA's Comprehensive Environmental Response, Compensation, and Liability Act, 42 U.S.C. §§ 9601 et seq. (CERCLA).

As a consequence of little prior planning for litigation support in the study design of the site investigation, data quality and sample traceability issues were uncovered in the soil samples. Once soil data issues had been addressed, a master soils PAH data set was constructed and various source analysis approaches were evaluated. Groundwater PAH data were plotted spatially but were not further analyzed. Use of parent PAH cross plots for soils data analyses were limited due to the exclusive selection of priority pollutant PAHs. Results of the cross plotting of 4- and 5-ring PAH ratios following Costa and Sauer.[15] showed wide ratio variability (Figure 1). The fluoranthene/pyrene ratio was both below 1.0 and above 1.2 in the master data set with significant middle ground demonstrated in a substantial number of samples and proved to be of little use in distinguishing among possible sources, e.g., urban background, MGP-derived coal tar residues and petroleum spills.[5] Attempts to relate any potential data groupings to one another could have only been done in the most arbitrary manner and would have been subject to legitimate and immediate legal challenge. If better discrimination among samples were to result, it was apparent that data analysis should avoid double ratio plots and move to other analyses that would include all PAH data.

First the overall hydrocarbon distribution of the entire PAH data set was examined on an analyte by analyte basis. Results showed that some PAHs were normally distributed, showed a bimodal distribution (a term used for a combination of a normal distribution mixed together with an appreciable number of non-detected values) and still others were skewed to low or non-detected values. Individual PAH concentrations were divided by total PAH ($\Sigma$ PAH17), as the 16 EPA priority pollutants PAHs were included in the dataset as was 2-methylnaphthalene, to normalize the entire dataset. Why 2-methylnaphthalene was in the

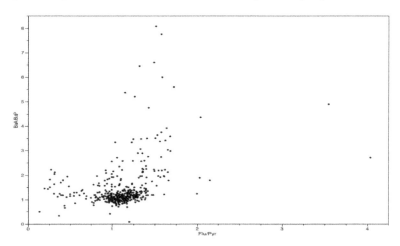

**Figure 1**   *X-Y Double PAH Ratio Source Plot of Fluoranthene/Pyrene (Flu/Pyr) by Benz(a)anthracene/Benzo(a)pyrene (BaA/BaP) for Study Case No. 1 Showing Wide Data Variability in Both PAH Source Ratios*

target analyte list was unclear. Normalization also assists in reducing laboratory specific analytical quantification differences among samples. For the normalized PAH dataset, principal component analysis (PCA) was selected as a viable data analysis approach following Johnson et al.[19] In PCA, absolute concentration PAH data (i.e., not normalized) should not be analyzed to avoid concentration driven and, hence, artificially skewed results. Consequently, only the Σ PAH17 normalized master dataset was analyzed by PCA on correlations. Not detected values were set to zero for statistical analysis.

The master dataset was analyzed using JMP statistical visualization software (SAS JMP version 9.0). The first three principal components were calculated, covering 67% of the combined variability for the entire dataset. A two dimensional X-Y plot of PC1 and PC2 showed some discrimination among sample points (Figure 2). However, a three-dimensional data plot using the first three principal components (Figure 3) yielded additional, more valuable, data discrimination. The three-dimensional data display using the first three principal components showed a semi-triangular data display consistent with the presence of at least three separate source components in the PAH data. End member data points from the three ends of the overall data distribution were sub-selected and locations compared with the overall coal tar reprocessing engineering diagram. The first grouping (colored green in Figure 3 for clarity) was completely located within the raw product storage area of the former coal tar processing facility. Locations of the samples in the other two groups (colored gold and purple in Figure 3) were not confined to any single area of the former facility. Individual distributional plots of the three end member soils groupings were plotted (Figures 4-6). Results showed that the raw product storage area (green grouping) was dominated by higher molecular weight PAHs, ranging from anthracenes and pyrenes to fluoranthene, benzofluoranthenes and perylene. Naphthalene was virtually absent from the higher molecular weight green soils grouping (Figure 5). For the gold soils grouping, naphthalene dominated

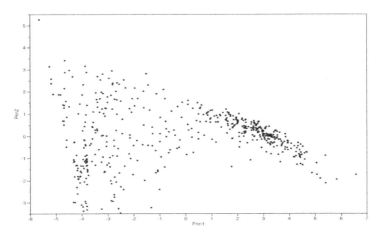

**Figure 2**   *X-Y Plot of PC1 and PC2 of All PAH Data Showing Some Discrimination Among Sample Points for Case Study No. 1*

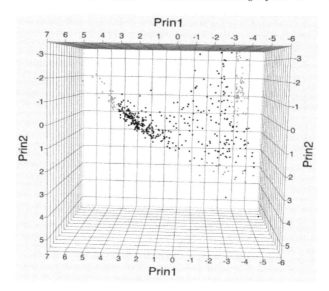

**Figure 3**   *X-Y-Z Plot of First Three Principal Components (PC1, PC2 and PC3) for All PAH Data in Case Study No. 1 Showing Group Discrimination Tied to PAH Source Differences*

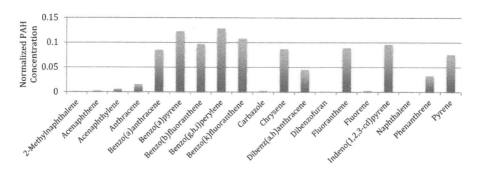

**Figure 4**   *The PAH Distributional Plot of a Representative Sample Selected from the Green End Member Grouping Showing Dominance by 4-and 5-ring Pyrogenic PAHs for Case Study No. 1*

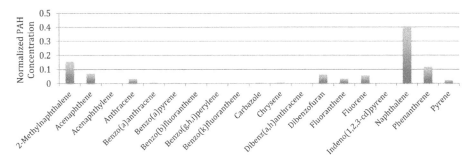

**Figure 5**  *The PAH Distributional Plot of a Representative Sample Selected from the Gold End Member Grouping Showing Naphthalene Dominance for Case Study No. 1*

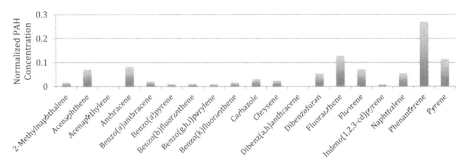

**Figure 6**  *The PAH Distributional Plot of a Representative Sample Selected from the Purple End Member Grouping Showing Phenanthrene Dominance for Case Study No. 1*

with sample points distributed throughout the site, from the raw product incoming storage areas to product manufacturing areas. For the purple soils grouping, phenanthrene dominated the PAH distribution, again with sample points being distributed throughout the site (Figure 6). For groundwater samples, naphthalene was found to dominate the groundwater dataset with highest concentrations clustered around and somewhat down gradient from incoming product storage areas. High soils naphthalene samples were not coincident with high groundwater naphthalene samples, indicating solubilization/soils washing, offsite transport or perhaps product compositional effects for naphthalene in groundwater.

For Case Study No. 1, coupling the PCA plots with PAH distributional plots served to illustrate the residual impacts of MGP and coal tar processing conducted over decades at the subject site. Double ratio plot analysis results proved not to be as compositionally sensitive and did not provide comparable sample discrimination to that of PCA. For Case Study No. 1, having not included 2- and 3-ring PAHs such as naphthalene, phenanthrene and other lower molecular weight PAHs in the dataset would have produced results that did not adequately capture the effects of MGP operations and coal tar reprocessing at and in the vicinity of the subject site. Focusing on 4-and 5-ring PAH analytes in double ratio plots would have seriously underestimated the net environmental impacts of former operations and any determination of associated remediation cost allocations would have been in error. Including data on lower molecular weight PAHs such as naphthalene and phenanthrene provided the key to source allocation that was consistent with the engineering diagrams from the MGP plant facilities that long since had been removed.

However, simply doing PCA analysis of the PAH data without development of an understanding of why the different data groupings were obtained was not helpful either. A better understanding of differential site contamination was obtained only when the PAH composition of each end member of the PCA three dimensional plot was coupled information on what site-based engineering processes were carried out where. Spatial restriction of one pyrogenic derived PAH distribution was key to sorting out legacy contamination sources for the assignment.

## 4.2 Source Analysis – Case Study Number 2

Case Study Number 2 involved a dispute over relic PAH contamination from historical MGP operations in the municipality that had operated in U.S. New England. The main dispute was coverage of cleanup costs from either the town taxes as represented by the town council or the separate non-tax related income from the municipal utility as represented by a separate utility board. The MGP site was not in a marine or coastal location, but did border on a fresh water municipal water supply. The municipality was facing significant soil and sediment PAH cleanup costs on publically owned property that the town council believed was caused by historical operations of a coal gasification derived electrical utility that was the predecessor of the current municipal utility. The municipality hired a consulting company that proceeded to collect soil and sediment samples, coordinate chemical analyses and prepare technical reports. Soil and sediment samples were analyzed for U.S. EPA priority pollutant PAHs plus 2-methylnaphthalene. Only after meeting regulatory deadlines did the town council turn to forensic efforts to support litigation against the municipal utility for cost recovery for investigation, legal, and planned remediation costs.

A total of 60 soil and fresh water sediment samples were collected and analyzed for the target PAHs. Laboratory PAH data were analyzed using ratio approach (Figure 7) and it was concluded that pyrogenic hydrocarbons dominated. It was concluded that results showed that the PAH site contamination was from relic MGP wastes. As the fluoranthene/pyrene versus benzo(a)anthracene/benzo(a)pyrene plot shows, there was a great deal of variability among sample results. Based upon the Figure 7 data, selection of potential source versus impacted areas was difficult. Results showed that concentrating only on a pyrogenic PAHs ratio data analysis approach yielded a biased picture of the site PAH contamination.

Statistical analysis of the entire PAH dataset revealed that individual PAH analytes were either skewed to low or non-detected values or approximated a normal distribution. Individual PAH concentrations were normalized by dividing by total PAH ($\Sigma$ PAH17) and the entire dataset was reanalyzed by using JMP statistical visualization software using correlations. Not detected values were set to zero for statistical analysis. The first three principal components (PC) accounted for 78% of the data variability and a three dimensional plot of the PC values revealed a complex mixing scenario of separately sourced PAHs (Figure 8). For clarity, sample locations were color-coded (blue for sediment, green for general site surface soils and red for soils near the former MGP plant). Based upon the PCA results, separate areas of differing PAH distributions were identified in the three groups. In the aquatic sediment samples (blue) two PAH distributions were identified: the first was pyrogenic in nature (Figure 9) and the second was almost exclusively dominated by naphthalene (Figure 10). In the middle of the PCA plot where it might have been expected to have mixing between the naphthalene dominated source and the pyrogenic source, a different PAH distribution was found (Figure 11) composed mostly of 2- and 3-ring PAHs consistent with a petroleum derived signature. Re-review of gas chromatograms collected by the engineering company showed an unresolved complex mixture (UCM) characteristic of petroleum was present in some soil and sediment samples. A petroleum derived PAH source had not been identified previously. Further historical research identified a "town dump" area within the MPG impact area that had received petroleum waste consistent with middle distillate fuels.

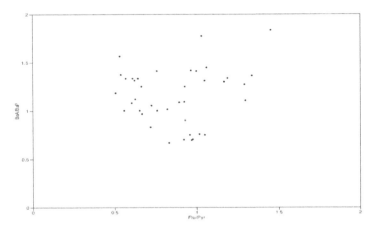

**Figure 7**   *X-Y Double PAH Ratio Source Plot of Fluoranthene/Pyrene (Flu/Pyr) by Benz(a)anthracene/Benzo(a)pyrene (BaA/BaP) for Case Study No. 2 Showing Data Variability in Both PAH Source Ratios*

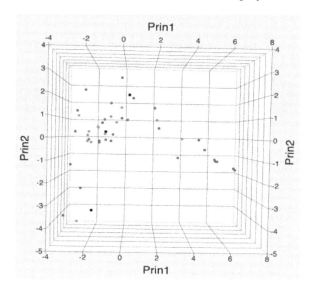

**Figure 8**  X-Y-Z Plot of First Three Principal Components (PC1, PC2 and PC3) for All PAH Data in Case Study No. 2 Showing a Complex Mixing Scenario of Separate PAH Sources

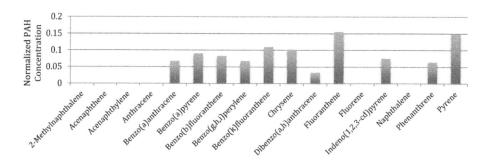

**Figure 9**  The PAH Distributional Plot of a Representative Sample Selected as the Blue End Member Grouping (Sediment) Showing Dominance by 4-and 5-ring Pyrogenic PAHs for Case Study No. 2

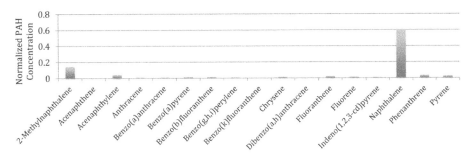

**Figure 10** *The PAH Distributional Plot of a Representative Sample Selected as the Red End Member Grouping (MGP Plant Related) Showing Dominance by Naphthalene for Case Study No. 2.*

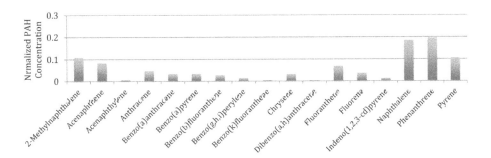

**Figure 11** *The PAH Distributional Plot of a Representative Sample Selected as the Green End Member for the Middle Mixing Area Grouping Showing the Presence of Petroleum-related PAH Distribution in the for Case Study No. 2*

The three sources identified in the statistical analysis of Case Study No. 2 PAH data were pyrogenic coal tar, naphthalene scrubber waste and a middle distillate petroleum product. Concentration on only pyrogenic ratio analysis such as fluoranthene/pyrene versus benzo(a)anthracene/benzo(a)pyrene did not provide information on two of the three sources eventually identified. Hence, reliance on PCA of the PAH dataset coupled with development of a more complete understanding of the historical and chemical signature of various PAH end member distributions resolved allocation issues between the town council and the town utility.

**4.3 Source Analysis – Case Study Number 3**

Case Study Number 3 stands in stark contrast to the first two cases. While Case Study Number 1 had hundreds of samples, limiting the PAH analyte list to the U.S. EPA Priority Pollutant PAHs (plus 2-methylnaphthalene) may have reduced the precision of the overall data analysis effort simply by data crowding. Case Study Number 2 had a smaller number of samples plus use of the U.S. EPA Priority Pollutant PAHs, so the possibility of data being overridden by other sample data points was reduced and better definition among sources might have been expected. However, again, with the limited PAH analyte list of the 16 Priority Pollutant PAHs (plus 2-methylnaphthalene), source definition was not straightforward. Case Study Number 3 on the other hand employed an expanded PAH analyte list of parent PAHs and their associated alkyl homologues as both coal gasification as well as petroleum product PAH contamination was anticipated. Moreover, in addition the U.S. EPA priority pollutant PAHs were analyzed for a number of vertical sediment profiles sampled in a suspected impact area immediately contiguous to the site.

The site for Case Study Number 3 was the industrial harbor area of a major United States coastal city. A MGP plant used to produce chemical manufacturing feedstock and generate electricity for the surrounding communities had been located at the site since the late nineteenth century. For decades the MGP plant fed a coal tar reprocessing facility that was located immediately next to it. Previous work had identified areas of known groundwater and soil contamination, active "fluid seeps" and contaminated coastal marine sediments in the immediate vicinity of the two sites. Governmental enforcement actions were causing the two successor companies to deal with legacy PAH contamination from the coal gasification, coke and coal tar reprocessing units at a single combined site.

Groundwater, soil, seep, tank contents and sediment samples were collected and analyzed for a suite of parent PAHs and associated alkyl PAHs. Parent and alkyl PAH data were normalized by dividing by total PAH ($\Sigma$ PAH39). Not detected values were set to zero for statistical analysis. Multivariate statistical analyses were completed on the normalized parent and alkyl PAH data using JMP statistical visualization software, including principal component analysis (PCA) of the combined suite of parent and alkyl homologue PAHs. PCA results showed that two NAPL samples and contaminated groundwater were similar and the seep and sheen samples (red group) were similar as well (Figure 12). However, the PAH signature of the tank contents sample was determined not to be closely related to any of the product or sheen samples (Figure 12). The PAH distribution of sediment samples (blue group) collected from the immediate seep area was not related to the sheen or NAPL samples (Figures 13 and 14). PCA results using normalized PAH data also showed that sediment PAH distributions were not uniform, with three separate end members present. A simple ratio plot of Benz(a)anthracene/Benzo(a)pyrene versus fluoranthene/pyrene (Figure 15) showed two generally separate groupings, with most of the suspected "source samples" (red group) were related to only one of the two identified sediment groupings (blue group). Sediment samples related to the expected "source samples" were generally found to be in the location of known coal offloading to the MGP facility. A single soil sample (green) was associated with one sediment group but not the other.

# Statistics of Lost Historical Coal Tar PAH Contamination in Soils and Sediments 119

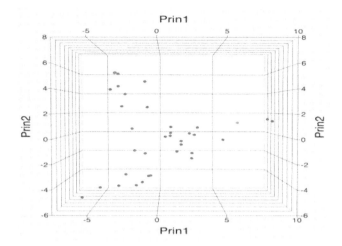

**Figure 12**  *X-Y-Z Plot of First Three Principal Components (PC1, PC2 and PC3) for All PAH and Alkyl Homologue PAH Data in Case Study No. 3 Showing Group Discrimination Tied to PAH Sources for Some Samples But Not for Others*

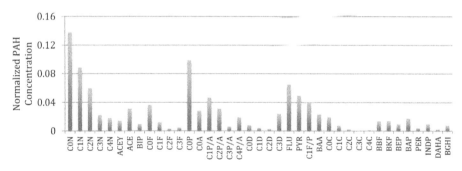

**Figure 13**  *PAH Distribution of Sample of Floating Non-aqueous Phase Liquid (NAPL) for Case Study No. 3*

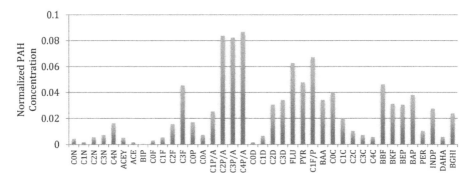

**Figure 14**  *PAH Distribution of Sediment Sample Selected for End Member Comparison with Floating NAPL for Case No. 3*

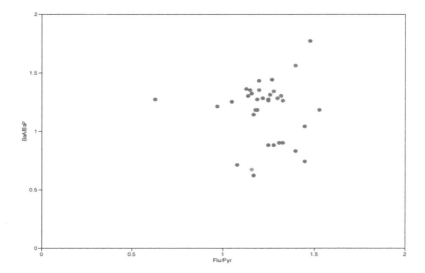

**Figure 15**  *X-Y Plot of Fluoranthene/Pyrene (Fly/Pyr) versus Benz(a)anthracene/ Benzo(a)pyrene (BaA/BaP) Showing Two Generally Separate Groupings, with Most of the Suspected "Source Samples" (Red Group) Related to Only One of Two Identified Sediment Groupings (Blue Group).*

Sediment vertical profile data provided a second approach to examination of PAH distributions in the sediment column resulting from operations at the MGP plant. Multiple vertical cores were collected and chemically analyzed for the U.S. EPA Priority Pollutant PAHs only. For statistical analysis the Priority Pollutant PAH data were not normalized; they were analyzed by least squares linear regression analysis to show concentration relationships within the vertical sediment column. To illustrate the results that were obtained, one sediment core was subsampled at four separate depths and U.S. EPA Priority Pollutant PAHs (without 2-methylnaphthalene) analyzed. When PAH concentrations in the bottom three samples were plotted against concentrations against the top sediment sample (Figures 16A, 16B and 16C), the data indicated enrichment of naphthalene deeper within the sediment when compared to other PAH analytes. Such enrichment could argue for a separate source of naphthalene contamination alone (i.e., MGP naphthalene scrubber discharge) unrelated to other PAH sources such as accidental releases of petroleum, urban background, or coal tar wastes. Results of the vertical sediment profiling in this case added to the cleanup area under consideration of the impacts of the MGP and coal tar processing impacts. With consideration only for the PAH ratio plot data analysis approach, the potential impacts of release of naphthalene scrubber waste at the MGP site may well have remained undiscovered.

5    DISCUSSION

Besides unwarranted exclusion of selected PAH from some data analysis approaches, there are three other issues that deserve further exploration: 1) the use of PCA analysis can provide an opportunity to calculate individual source contributions, 2) the evaluation of the consistency, or lack thereof, of the PCA pyrogenic PAH source signal across the three sites, and 3) the distribution of individual PAHs at sites compared across sites. First, use of PCA data to

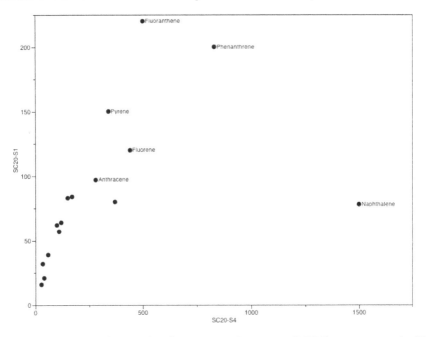

**Figure 16A**   *X-Y Plot of Vertical Sediment Data, Comparing PAH Concentrations (ug/Kg) in the Bottom Most Sediment (#4) Sample (X-Axis) with the Shallowest (#1) Sediment Sample (Y-Axis), Showing Significant PAH Enrichment in Sediment Samples from the Vertical Core (#20) for Case Study No. 3*

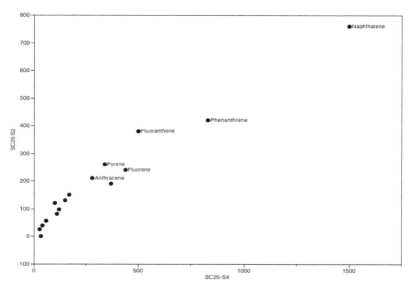

**Figure 16B**  *X-Y Plot of Vertical Sediment Data, Comparing PAH Concentrations (ug/Kg) in the Bottom Most Sediment (#4) Sample (X-Axis) with A Shallower (#2) Sediment Sample (Y-Axis), Showing Little PAH Enrichment in the Vertical Core Sediment (#20) for Case Study No. 3*

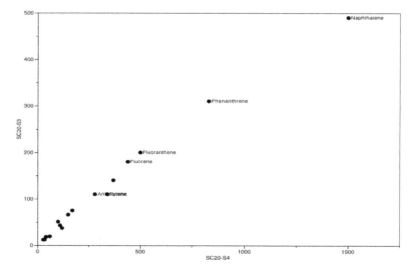

**Figure 16C**  *X-Y Plot of Vertical Sediment Data, Comparing PAH Concentrations (ug/Kg) in the Bottom Most Sediment (#4) Sample (X-Axis) with Next Shallower (#3) Sediment Sample (Y-Axis), Showing Little PAH Enrichment in the Two Deepest Samples from the Vertical Core (#20) for Case Study No. 3.*

calculate the amount of contribution at any single point from more than one PAH source signal could allow source apportionment among two or more principal responsible parties. Careful consideration of the details within PCA results is required, however. Second, reducing the number of PAH analyses reduces source discrimination power to an unspecified extent when only U.S. EPA Priority Pollutant PAHs are used, especially when compared to PAH and alkyl homologue based PAH datasets of the type given in Table 1. Such an effect may be illustrated (and perhaps quantified) when considering the details of pyrogenic PAH distributions for the three case studies discussed. Third, analysis of the population distribution of individual PAH compared across case studies revealed a consistency for some prominent pyrogenic PAHs but not for others at all three coal gasification sites. Further, population distributions for petrogenic PAH such as the two-ring and three-ring PAHs were not consistent when compared across all three sites; such a finding was consistent with the presence or lack of presence of petroleum products at each site. Naphthalene showed a skewed distribution for all three sites, a finding that perhaps was most consistent with how naphthalene was dealt with at each site over time.

## 5.1 Calculation of Source Mixing for Any PCA Sample Point

Having identified candidate "End Members" within the PCA distributional plot, it is possible to calculate the degree of mixing between two end members or among three end members for any sample point within a PCA distributional continuum. New PAH concentration data from the end members are calculated by mixing sample results stepwise starting with end members A/B at 100%/0% and proceeding every 10% to 0%/100%. Calculated data are added to the normalized PAH data set and the PCA analysis redone. For three end member mixing calculations, several sets of calculated data can be completed and the PCA analysis redone. To illustrate two-end member mixing behavior, Figure 17 shows a PAH dataset with the stepwise 10% mixing line added for emphasis between end members A and B. Results show that while the calculations are linearly proportional, PCA results are not. Figure 18 shows the same X-Y-Z Plot shown in Figure 17 with a second mixing model added using two different "End Members" in the same dataset. Results clearly show that the two different mixing models do not have the same internal behavior. Such behavior can only mean that results of mixing models of different PAH distributions have different proportional contributions to the final calculated sample. In mathematical terms, mixing model results appear to be signal strength (i.e., "end member") dependent. Simply assuming that a linear mixing of multiple PAHs gives a linear result regardless of proportion is not correct for PCA analysis. While percentage values for any single sample point falling along the mixing line can be calculated using the mixing model results, care must be paid to how each mixing model in each dataset behaves. Such modeling results can only be plotted to show mixing of sources at a site or be used to establish remediation financial responsibilities under the most carefully understood conditions. Simply assuming a given site-specific mixing model adequately calculates differences is not sufficient. Further, it must be recognized that mistakes in the selection of individual end members can prejudice results no matter what statistical analysis approach is employed to calculate individual contributions.

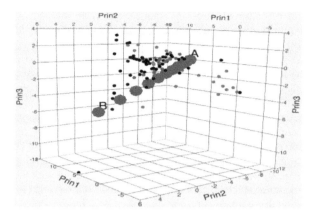

**Figure 17**  X-Y-Z Plot of the First Three Principal Components of a Soil and Sediment PAH Dataset Showing a Linear Mixing Model of Two PCA End Members That Resolves Itself Into Non-Linear Spacing on the PCA Plot

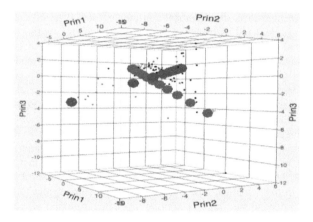

**Figure 18**  X-Y-Z Plot of the First Three Principal Components of a Soil and Sediment PAH Dataset Showing a Second Linear Mixing Model With Clearly Different Non-linear Spacing Compared with the First Mixing Model

### 5.2  Comparison of Pyrogenic PCA "End Members" Among All Three Case Studies

Without exception when studying MGP sites, reference samples of the PAH in coat tar or other waste stream materials indicative of operational contaminants do not exist. The simple passage of time works against retention of reference waste streams from the nineteenth and early twentieth centuries for twenty-first century forensic investigations. Therefore, there are no site-specific "PAH reference standards" by which impacts can be compared by forensic geochemical testing for source assignments. PCA allows selection of the next best available things by identifying the "end members" of the data distribution sub-groups found for each plot from each site. However, care using PCA results must be exercised. An interesting question regarding the chemical composition of MGP and coal tar waste might be to what degree of consistency is found for pyrogenic sources in multiple MGP and coal tar assessments over time and location for the three MGP Case Studies. Figure 19 provides a comparison of U.S. EPA Priority Pollutant PAH distributions from the three Case Studies discussed in this paper. In Figure 18, relative PAH data from Case Study 1 (Figure 19A) and 2 (Figure 19B) were re-plotted to exclude 2-methylnaphthalene and only the Priority Pollutant PAH data from Case Study 3 were plotted (Figure 19C). Individual PAH data were not renormalized to reflect the reduced number of PAHs in the distributional plots. Figure 18 distributional plots show that there is a measure of consistency among the three data Case Study pyrogenic plots but overall the plots are not completely identical. However, for all three Case Studies the ratio of fluoranthene to pyrene ranged from approximately 0.8 to slightly more than 1.0 in the pyrogenic "end members" selected from the PCA. Values were at or slightly higher than pyrogenic PAH distributions as discussed in the literature,[5] perhaps demonstrating differential weathering of pyrogenic source signal(s) over the decades since release. While the fluoranthene to pyrene ratio for the samples selected in all three Case Studies for the pyrogenic "end members" of the PCA distribution were below or close to 1.0, the PCA results themselves clearly reflect site to site differences for other PAH sources. However, when analyzing PAH datasets, reducing the number of PAH analytes in data analysis does not portend good outcomes. It may be that the pyrogenic end members within a U.S. EPA Priority Pollutant dataset may show some degree of consistency from MGP site to MGP site - perhaps by the very nature of the engineering processes involved and the reduced amount of information within a parent PAH (i.e., the U.S. EPA Priority Pollutant PAHs) dataset only - no such consistency on the presence of naphthalene or petroleum sourced PAHs at any individual site should be assumed. Site–to-site differences on the presence of naphthalene (e.g., how "naphthalene problems" were handled or did not have to be handled) should be expected to be present and, therefore, all PAH data examined statistically. Based upon such an outcome, reliance on individual PAH ratios using 4- and 5-ring PAHs only may not provide a complete picture of the extent of the MGP operational impacts for any single site.

## 5.3 The Degree of Data Distributional Consistency Among Individual PAHs Across Case Studies

The data distribution of the relative concentrations of individual PAHs was plotted for all three case studies (Table 2). There was no consistent data distribution among individual PAHs for the three studies: some PAHs were normally distributed, some PAHs where skewed to lower or non-detect values and some were bimodal. Case Study No. 1 had PAH (U.S. EPA Priority Pollutants plus 2-methylnaphthalene) data from 522 samples, Case Study No. 2 had PAH (U.S. EPA Priority Pollutants plus 2-methylnaphthalene) data for 49 samples and Case Study

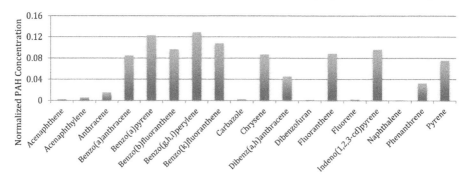

**Figure 19A**  *PAH Distribution of Case No. 1 Pyrogenic PCA End Member*

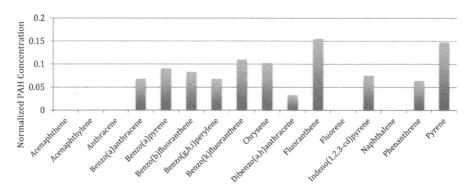

**Figure 19B**  *PAH Distribution of Case No. 2 Pyrogenic PCA End Member*

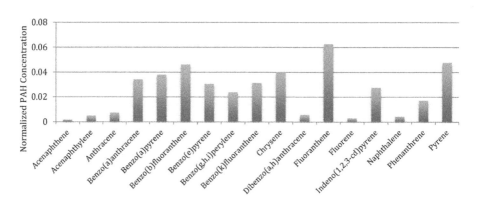

**Figure 19C**  *PAH Distribution of Case No. 3 Pyrogenic PCA End Member*

No. 3 had PAH (parent and alkyl homologues) data for 36 samples. Most of the combined distributional results (labeled bimodal in Table 2) were found in Case Study No. 1, the case with the most samples. The bimodal outcome might have been expected due to the larger number of samples compared to the other two case studies plus the lack of coherent forensic sampling strategy for Case Study No. 1. A normal PAH distribution across the three case studies was found only for anthracene and fluoranthene. However, a second kind of consistency, either bimodal (Case Study No. 1) or normal (Case Study No. 2 and 3) distributions were found for benzo(a)anthracene, benzo(a)pyrene, benzo(b)fluoranthene, and chrysene. The other PAHs were mostly skewed to lower levels or non-detected values. Naphthalene distributions were consistently skewed in all three case studies, but in Case Study No. 1 with extreme (the gold colored grouping) high relative concentrations (>50%) mixed in with lower relative concentrations. There were few to none for Case Studies No. 2 and 3, respectively. The higher molecular weight 4- and 5-ring PAHs (i.e., pyrogenic sourced PAHs) showed the most distributional consistency among the three case studies. Going forward, a working hypothesis might be that such consistency indicates high temperature process-derived PAHs in all three case studies; similarly, the lack of consistency for petrogenic PAHs and for naphthalene indicates a variable source effect across the three case studies in the final PAHs data distribution for petrogenic as well as single naphthalene (gasification waste) sources. Variable uses of petroleum products at the various sites as well as how individual facilities dealt with their "naphthalene problem" appear to be reflected in the distribution of such sourced PAHs.

**Table 2**  *Summary of Data Distributions of Individual PAHs for All Three Case Studies*

| PAH Analyte | Case Study No. 1 (N= 522) | Case Study No. 2 (N=49) | Case Study No. 3 (N=36) |
| --- | --- | --- | --- |
| Acenaphthene | skewed | skewed | skewed |
| Acenaphthylene | skewed | normal | skewed |
| Anthracene | normal | normal | normal |
| Benzo(a)anthracene | bimodal | normal | normal |
| Benzo(a)pyrene | bimodal | normal | normal |
| Benzo(b)fluoranthene | bimodal | skewed | normal |
| Benzo(g,h,i)perylene | skewed | skewed | normal |
| Benzo(k)fluoranthene | skewed | skewed | skewed |
| Chrysene | bimodal | normal | normal |
| Dibenzo(a,h)anthracene | bimodal | skewed | skewed |
| Fluoranthene | normal | normal | normal |
| Fluorene | skewed | skewed | skewed |
| Indeno(1,2,3-cd)pyrene | skewed | skewed | skewed |
| Naphthalene | skewed | skewed | skewed |
| Phenanthrene | normal | normal | skewed |
| Pyrene | normal | skewed | normal |

## 6. SUMMARY

Consideration of the selection of the PAH analyte list for investigation of historical MGP site operations is driven by any number of considerations including laboratory selection, analysis costs, data quality, sample result turnaround time, and perhaps even investigator preferences. It appears that incomplete consideration has been given in the MGP investigation to the impacts the selection of the PAH analyte list has on the data analysis methods that then become available to the investigation. Reliance on published PAH ratio analysis approaches may or may not provide useable results in another case. Simply assuming comparability of site conditions being investigated to those that have been discussed in the literature is wrong. There is no guarantee that analytical results from one site study will be comparable to those from another just because one is conveniently discussed in the published literature. Data analysis approaches wherein all available data are used to establish PAH contamination limits on the impacted environment must be employed. Consideration of data represented by PCA end members found in a data plot must be coupled with an understanding of the geochemical significance of each source signal as impacted by historical information on the process(es) used at the site in question. To do any less invites legitimate legal challenge.

It is incumbent upon a forensics investigator to balance sample collection design with a number of laboratory analytical approaches to ensure that useable data analysis results are obtained. Such conclusions are especially true when dealing with naphthalene impacts of MGP site operations as there is considerable evidence that naphthalene impacts may, in fact, be more wide ranging than higher molecular weight PAH impacts that have been considered previously. Naphthalene enrichment in soils and sediments from MGP impacts appears to be related to the manner in which naphthalene was dealt with in the coal gasification engineering process(es) during the latter part of the nineteenth century. It is wrong to assume that each and every MGP facility handled the "naphthalene problem" in a manner identical to every other MGP facility. From a review of the nineteenth century literature on the subject, it is clear that just the opposite was true, that different facilities had different degrees of success in dealing with the "naphthalene problem." Some facilities did not report a "naphthalene problem" at all. Further, the presence or absence of petroleum derived PAH source(s) should be proven through an analysis of the PAH distribution, if present. There is no guarantee that significant petroleum sources can be assumed to be either present or absent based upon data from one PAH. Therefore, rigorous statistical data analysis of all PAH analytes coupled with geochemical interpretation of the results is required to quantify the complete suite of operational impacts of MGP operations as well as those from other PAH sources.

## References

1. Hatheway, A.W. 2006. World history of manufactured gas: a 'world' of land redevelopment possibilities. Proceedings of the International symposium, and Exhibition of the Redevelopment of Manufactured Gas Plant Sites 4-6 April 2006, Reading, UK. Published in Land Contamination & Reclamation 14(2). 2006.
2. Murphy, B. L., Sparacio, T., and Shields, W. J. 2005. Manufactured gas plants-processes, historical development, and key issues in insurance coverage disputes. Environmental Forensics 6:161–173.

3   Hamper, M.J. 2006. Manufactured Gas History and Processes. Environmental Forensics 7:55–64.
4   Saber, D., Mauro, D., Sirivedhin, T. 2006. Environmental Forensics Investigation in Sediments near a Former Manufactured Gas Plant Site. Environmental Forensics 7:65–75.
5   Boehm, P.B. 2006. Chapter 15. Polycyclic Aromatic Hydrocarbons (PAHS). In Environmental Forensics. Contaminant Specific Guide. Edited by R.D. Morrison and B.L. Murphy. Elsevier, Inc. Academic Press. London. 541 pp.
6   Blumer, M. 1977. Polycyclic aromatic compounds in nature. Scientific American. 234(3) 34-45.
7   Neff, J.M. 1979. Polycyclic aromatic hydrocarbons in the aquatic environment. Sources, fates and biological effects. Applied Science Publishers Ltd. Essex, England. 261 pp.
8   Neff, J.M., Ostazeski, S.A., Macomber, S.C., Roberts, L.G., Gardner, W. and Word, J.Q. 1998. Weathering, Chemical Composition and Toxicity of four western Australian crude oils, Report to Apache Energy, Ltd., Perth, Western Australia, Australia.
9   Herzog, I. 1874. On naphthalene. Proceedings of the American Gas Light Association. Vol. I, page 66. May 1874.
10  Hyde, G.A. 1874. Naphthalene. Proceedings of the American Gas Light Association. Vol. I, page 68. May 1874.
11  Curley 1877 Naphthalene. Proceedings of the American Gas Light Association, Volume III, Page 67
12  Dresser 1877 Naphthalene. Proceedings of the American Gas Light Association, Volume III, Page 70
13  Walker, J.H. 1885. Naphthaline. Proceedings of the American Gas Light Association. Vol. VII, page 157. October 1885.
14  Harper, G.H. 1894. Applications of chemical technology to gas manufacture. Proceedings of the American Gas Light Association, Vol. XI, No. I. Pages 149-168. July 1894.
15  Progressive Age. Gas-Electricity-Water. 1891. Vol. 9. No. 6. New York. March 16, 1891. Page 121.
16  Progressive Age. Gas-Electricity-Water. 1891. Vol. 9. No. 7. New York. April 1, 1891. Page 140.
17  Costa, H.J. and Sauer, T.C. 2005. Forensic approaches and considerations in identifying PAH background. Environmental Forensics. 6:9-16.
18  Herman, K.D., Wannamaker, E.J., Jegadeesan, G.B. 2012. Sediment PAH allocation using parent PAH proportions and a least root mean squares mixing model. Environmental Forensics 13:225-237.
19  Johnson, G.W., Ehrlich, R., Full, R. 2002. Principal components analysis and receptor models in environmental forensics. Chapter 12. In: Introduction to Environmental Forensics. Edited by B.L. Murphy and R.D. Morrison. Academic Press. San Diego. 560 pp.

# MICRO-BUBBLES OXYGEN INJECTION IN GROUNDWATER CONTAMINATED BY ORGANIC BIODEGRADABLE COMPOUNDS AND METALS

A. D'Anna[1], R. Brutti[1], A. Gigliuto[1], R. Vaccari[1], G. Bissolotti[2], E. Pasinetti[2] and M. Peroni[2]

[1] AECOM Italy srl, Via F. Ferrucci 17/A, 20145, Milano, Italy
[2] SIAD S.p.A., Laboratorio di Biologia e Chimica Ambientale, Via Pasubio 5, 24040 Dalmine, Bergamo, Italy

## 1 INTRODUCTION

In this paper we present the results obtained in an in-situ micro-bubble injection of pure oxygen into the groundwater (GROUND $BIO_2^®$ system developed by SIAD S.p.A.), carried out in an active plant of a pharmaceutical company in northern Italy. The objective of the test was to evaluate the feasibility of a bio-stimulation technology, with pure oxygen, for the remediation of groundwater contaminated primarily by petroleum hydrocarbons and aromatic organic compounds (BTEX).

The test involved the injection in a monitoring well of a mixture of oxygen and nobles gases in order to highlight the effectiveness of removal of contamination and identify the site-specific radius of influence (ROI) of the technology according to the site specific hydrogeology.

Subsoil is characterized by the presence of unsaturated soil mainly consisting of gravel - sandy silt matrix from ground level down to a depth of about 10 - 15 m b.g.; from about 10 m to 15-20 m b.g.l., the saturated soil consist mainly of silty-sands to sandy-silts; from 18,5 to 21.m b.g.l. (max depth reached) there is a layer of clay (thickness of 3 - 6 m). Periodic groundwater monitoring , define an average depth to groundwater levels at the site of approximately 9,5 m; groundwater flow direction is from NW to SE, with an hydraulic gradient of between 5‰ and 8‰. Hydrogeological tests conducted on the site indicated a hydraulic conductivity of 6,83 x $10^{-5}$ m/s, indicating an aquifer characterized by a moderate permeability.

## 2 MATERIAL AND METHOD

### 2.1 Technology description

Micro-bubbles pure oxygen injection technology used (GROUND $BIO_2^{(R)}$) and noble gases mixed with oxygen (GROUND $MIX^{(R)}$) are patented by SIAD S.p.A. They are based on the pressure injection of gas in groundwater, through a diffuser with antifouling microporous membrane with a high efficiency of gas dissolution. The supply of pure oxygen allows high concentrations of dissolved oxygen to be reached in the injection point

(40 - 50 mgO$_2$/L @ atmospheric pressure, increasing with increasing of water head above diffuser). These high concentration sustain the oxygen diffusion in the neighbourhood, due to the concentration gradient (molecular diffusion) created. The reached concentration are more than four times greater than with traditional systems (like AirSparging – AS or Biosparging - BS, etc.).

Dissolved oxygen stimulates the aerobic processes for the degradation of organic compounds in groundwater, promoted by the indigenous microorganisms of the site.

The principle of diffusion for concentration gradient along with low flows injection (equal to 1 – 1,5 NL/h), minimizes or eliminates the volatilization phenomena of the contaminants (no stripping effect).

The injection system is very compact and also has no electric part so the system is practically an Explosion proof (Ex) system; the pressure of cylinder is the "engine" that drives oxygen into the diffuser / piezometers.

The use of noble gases such as neon and krypton, allows us to easily define the ROI of the technology (in time and distance) in the first month (noble gases are not subject to biological and chemical consumption but have a similar GW diffusion as oxygen).

## 2.2 Field Scale test description

Prior to field scale test we perform a lab scale test to evaluate, under controlled conditions (in batch), the feasibility of bioremediation by injection of pure oxygen in site specific water. The Lab test showed the complete water removal of BTEX and Hydrocarbons in 5 months and Chemical analysis of batch head air (by to Activated Carbon) excluded stripping of contaminants.

The in-situ tests began in 2012 and lasted for 17 months. The test involved the injection of a mixture of pure oxygen and noble gases (neon and krypton) in a contaminated well (P1). This test was to assess the response of the injection wells and surrounding areas:

- the effectiveness of bioremediation by indigenous biomass as a result of stimulation with pure oxygen,
- the site-specific radius of influence.

The wells used for the field test were: injection (P1), one monitoring well located 3 meters from the injection well perpendicular to GW direction (P2), one monitoring well located 5 meters from the injection well perpendicular to GW direction (P3) and one monitoring well located approximately 25 meters downgradient of injection well along the main GW direction (P4).

During the test monthly monitoring was performed for the following parameters:
- hydrochemical parameters,
- analysis of contaminants,
- analysis of noble gases,
- microbiological specific analysis.

In addition to the chemical analysis as shown in the course of the tests microbiological analyzes were carried out in order to verify that the injection of pure oxygen would promote contaminant specific growth in the groundwater. In particular, bacterial counts were performed and total bacterial counts (aerobic and anaerobic) and for the specific strains suitable for the aerobic degradation of contaminants (hexane oxidants, toluene oxidants, etc.).

## 3 RESULTS

### 3.1 Site Contamination

The field test area analysis show that in the area there was high concentration of total hydrocarbon aromatics compounds (BTEX, mainly toluene), respectively with concentrations equal to 95,000 µg/L and 88,000 µg/L in the injection well P1; 61,000 µg/L and 56,000 µg/L in the P2 well; 9,000 - 10,000 µg/L and 8,000 µg/L in P3 and P4 wells. For aromatic compounds (BTEX) in addition to toluene are present with lower concentrations of benzene, ethylbenzene and xylene. Furthermore, due to the reducing condition in GW are present also iron II (from 5,000 µg/L to 32,000 µg/L) and manganese (4,000 µg/L and 9,000 µg/L).

### 3.2 Contamination trends

The below graphs shown the concentration trends of the total hydrocarbons and BTEX in the injection well and in the monitoring wells during the first 8 month of the test (the colours of trend/wells are shown in the right image). From the graphs we can see that the concentrations of biodegradable organic compounds have gradually decreasing to reach in about 8 months removal rates close to 100% throughout the field test.

### 3.3 Field analysis

The in situ tests include the hydrochemical parameters monitoring by multi-parametric probe. In particular were monitored: temperature, pH, conductivity, red-ox potential and dissolved oxygen concentration.

Among the physic-chemical parameters and hydrochemical detected during the test, dissolved oxygen and red-ox potential were the most important. In particular in the injection well (P1) dissolved oxygen increased from values equal to or below 2 mg/L to values higher than 30 - 35 mg/L (with a maximum of 45 - 50 mg/L) and the red-ox from typically reducing conditions (rH = - 168 mV), reached positive values typical of an oxidizing status (rH = + 40 mV).

In lateral and down gradient wells associated with contaminant concentrations reduction, was observed an increase of the red-ox potential, more close to oxidizing condition, while the dissolved oxygen from values lower to 0,2 - 1 mg/L is gradually increased up to values of about 5 mg/L.

In Figure 2 is notable the delay of the biodegradation start which is 2 months for the 3m lateral well and about 4 months for the lateral 5 meter well and for the 25 m downgradient well.

### 3.4 Noble gas analysis

In order to assess the radius of influence, for the first 4 months we injected a mixture of oxygen and noble gases (neon and krypton). The following figure 3 shows the concentration vs time of neon and krypton in the three monitoring wells (P2, P3 and P4).

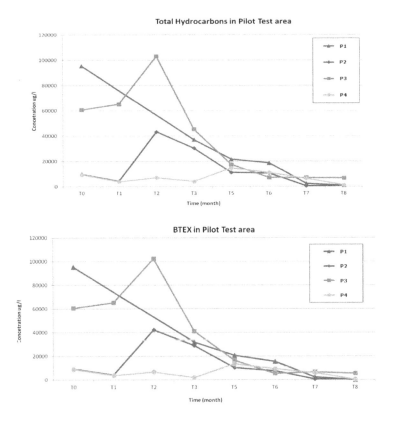

**Figure 1**  *Contaminants (Hydrocarbons and Aromatic compounds) trend in field monitoring wells (on x-axis Time in months, on y-axis concentration in µg/L)*

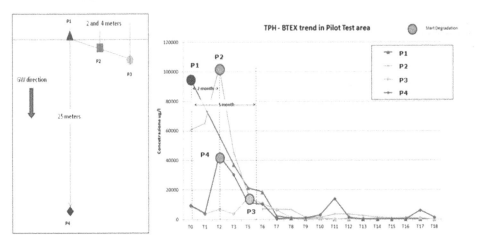

**Figure 2**  *Contaminants trend and starting of biodegradation process*

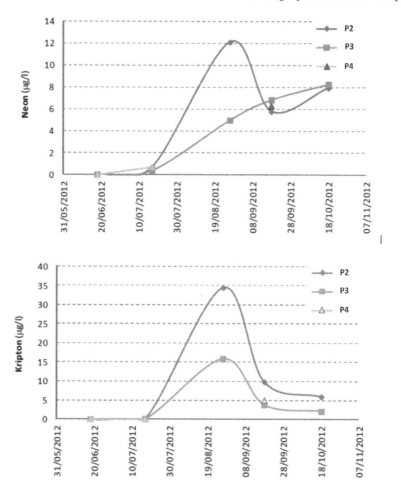

**Figure 3**   *Noble gases analysis (time vs concentration of noble gases)*

As can be seen from the data and graphs, both noble gases have been found in all the monitoring wells. In particular, we found-it in P2 and P4 after about 2 months from the start of injection, while recording higher concentrations in P2 well. For the P3 well noble gases was observed after 3 months from the start of injection. This diffusion is slow but reaches the surrounding of the injection well with a lateral radius of influence at least 5 meters; in the downgradient area we note the effect at least 30 meters.

So the site-specific application of technology shows (with low concentration) a radius of influence of about 8 to 9 m/month along the GW direction and of about 2 m/month laterally. The results obtained with the noble gases were confirmed by the contaminant removal, the increase of the concentration of oxygen and the growth of bacterial counts.

## 3.5 Microbiological analysis

The bacterial counts baseline values were equal to $10^{+1}$ - $10^{+2}$ MPN/mL in the injection well (P1), $10^{+2}$ - $10^{+3}$ MPN/mL in P2 and $10^{+3}$ MPN/mL in P3. During time the pure oxygen stimulation has supported bacterial growth with significant increases of more than four orders of magnitude in the P1 (up to $10^{+6}$ MPN/mL), more than two orders of magnitude in the P2 (up to $10^{+5}$ MPN/mL) and over an order of magnitude in P3. It is also noted as the oxidizing species were all stimulated by treatment with significant increases especially for hexane and toluene oxidizing strains.

The Bacterial concentration trends are correlated to the initial concentrations of contaminants in each well and their biodegradation activities during the field test. The increase of the oxidizing bacterial species, corresponds to a significant decrease in the total hydrocarbon contamination and toluene, with yields decreasing in function of the distance from the point of injection, in particular: removal rate close to 100% in the P1 (injection well - see graphs of Figure 5), removal rate of about 90% in P2 (3 meters lateral from injection well) and removal rate of 45% and 70% respectively for the total hydrocarbons and toluene, in P3 (5 meters lateral from injection well).

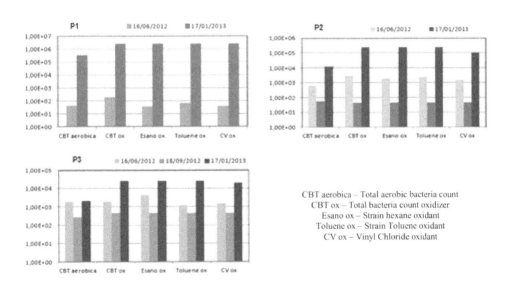

CBT aerobica – Total aerobic bacteria count
CBT ox – Total bacteria count oxidizer
Esano ox – Strain hexane oxidant
Toluene ox – Strain Toluene oxidant
CV ox – Vinyl Chloride oxidant

**Figure 4**   *Microbiological analysis data for baseline and after 6 months*

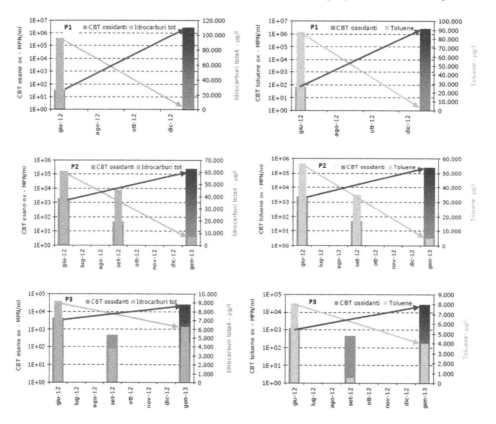

**Figure 5** *Bacterial counts (blue colour) vs contaminants concentration (Hydrocarbons in green and Aromatic compounds in orange)*

4 CONCLUSION

The test performed shows the results of an in-situ field test with micro-bubbles injection of pure oxygen into groundwater (GROUND $BIO_2^{(R)}$), within an active pharmaceutical factory in northern Italy. The objective of the test was to evaluate the feasibility of a bio-stimulation technology, with pure oxygen, for the remediation of groundwater contaminated primarily by petroleum hydrocarbons and aromatic organic compounds (BTEX).

It was initially evaluated with success the radius of influence of the technology by injection of a mixture of oxygen and noble gases.

The aerobic biodegradation was verified over time by monitoring the contaminant concentrations, physical-chemical parameters, hydro-chemical data and bacterial counts. The field test analysis showed that in the area there were high concentrations of total hydrocarbon aromatics compounds (BTEX, mainly toluene), respectively with concentrations equal to 95.000 µg/L and 88.000 µg/L in the injection well P1; 61.000 µg/L and 56.000 µg/L in the P2 well; 9.000 - 10.000 µg/L and 8.000 µg/L in P3 and P4 wells.

The noble gases analysis in the first months of the test showed that the diffusion of the gas reached the surrounding area with a radius of at least 4 to 5 meters in a lateral direction. In the downgradient area we note the effect at least 25 to 30 meters from the injection well.

During the test the concentration of biodegradable organic compounds gradually decreased to reach in about 8 - 9 months, removal rates close to 100%, both in lateral and in downgradient wells.

The bio-stimulation activity was also confirmed by microbiological analysis that shows that the autochthon bacterial biomass grew more than four orders of magnitude in the injection well and more than two orders of magnitude in the lateral wells. It was also observed that the oxidizing specific species (hexane and Toluene oxidizing bacteria were present ($10^{+6}$ MPN/mL). The growth of microorganisms was proportional to the reduction of chemical contamination: the increase in oxidizing bacterial species corresponded to a significant decrease in total hydrocarbons and BTEX contamination, with yields decreasing with the distance from the injection point.

We also report the following evidence about chemical-physical and hydrochemical parameters: in the injection well dissolved oxygen increased from values equal to or below 2 mg/L to values higher to 30 mg/L, with a decreasing trend from depth to the surface, and the red-ox potential by typical reducing conditions (rH equal to -168 mV), reached positive values typical of an oxidizing environment (rH equal to + 40 mV). Based on the considerations summarized above, we can say that the technology has a radius of influence of approximately 8 - 9 m / month in the GW direction and about 2 m/month of lateral diffusion.

Overall, activity in the in-situ testing showed that the chosen technology stimulates the biomass of autochthonous bacteria and thus promotes the aerobic removal of organic contaminants (hydrocarbons and BTEX) with removal rates close to 100%.

HEAVY METAL CONTAMINATION IN SELECTED ELECTRONIC WASTE DUMPSITES IN LAGOS, NIGERIA.

V. F. Doherty,[1] M. K. Ladipo,[2] and I. O. Famodun.[1]

[1] Environmental Biology unit, Department of Biological Science. Yaba College of Technology, Lagos.
[2] Department of Polymer and Textile Technology. Yaba College of Technology, Lagos.

1  INTRODUCTION

Electronic waste or e-waste refers to discarded electrical and electronic equipment. Any electrical or electronic equipment that is taken to no longer be capable of performing the function for which it was originally intended is an e-waste.[1] The useful life of consumer electronic products is relatively short, and decreasing as a result of rapid changes in equipment features and capabilities.[2] This creates a large waste stream of obsolete electronic equipment. Due to their hazardous material contents, Waste Electrical and Electronic Equipment (WEEE) may cause environmental problems during the waste management phase if it is not properly pre-treated.[3] The composition of e-waste is variable but contains more than 1000 different substances categorized as hazardous and nonhazardous and contains ferrous and nonferrous metals, plastics, glass, wood and plywood, printed circuit boards, concrete and ceramics, rubber and other items [3]. The presence of elements like lead, mercury, arsenic, cadmium, selenium and hexavalent chromium and flame retardants beyond threshold quantities in WEEE/e-waste implicates them as hazardous waste.[3]

Activities concerning electronic waste (e-waste) are emerging as a global concern as they can contribute to the release of heavy metal and persistent toxic substances (PTSs) into our environment. The disposal, recycling, and part salvaging of discarded electronic devices such as computers, printers, televisions, and toys are now creating a new set of waste problems.[4] Today, waste electrical and electronic equipment (WEEE) or electronic waste (e-waste) generation, trans-boundary movement and disposal are becoming issues of concern to solid waste management professionals, environmentalists, international agencies and governments around the world.[5,6] E-waste disposal is especially problematic when humans and the environment are exposed to hazardous chemicals during the process of dismantling electronic products.[7] With today's technologically advancing societies and the demand for newer, more efficient, and effective technology, older and outdated electronic items are becoming obsolete and are being discarded in significant amounts in various parts of the world.[3]

Nigeria has emerged the most important African importing country for new and used electronic and electrical equipment (EEE), which are shipped largely from the United Kingdom (UK).[8] In Nigeria, the consumption of electrical and electronic devices is increasing rapidly, which is leading to rapidly growing e-waste volumes. In Nigeria, many e-waste fractions cannot be managed appropriately, which is resulting in the accumulation of large hazardous waste volumes in and around major refurbishing and recycling centers. Furthermore, some recycling practices – like the open burning of cables and plastic parts – cause severe emissions of pollutants such as heavy metals and dioxins.[9]

According to a United Nations Environment Programmed (UNEP) report, an analysis of 176 containers of two categories of used EEE imported into the country, conducted from March to July 2010, revealed that more than 75 percent of all containers came from Europe, approximately 15 percent from Asia, five percent from African ports (mainly Morocco) and five percent from North America. A considerable amount of these imports, most of which are hazardous, end up at the Ikeja Computer Village in Lagos, believed to be the largest single market for computers and allied products in the West African sub-region.[10] Of all the items traded, only 25% can be recycled or reused, the remaining 75% scrap becomes a mountain of environmental toxicity, for many developing nations.[11]

The aim of this study is to determine the concentration of heavy metals (Pb, Zn, Cd, Ni, Cr and Cu) in soil and water samples collected from and around electronic waste dumpsites in different parts of Lagos state (computer village (Ikeja), Alaba International Market (Ojo) and Westminister (Apapa).

2 METHODOLOGY

**2.1 Study area**

The study area is situated in Lagos State, Nigeria. Lagos State is located in the southwest part of Nigeria in West Africa. It is on coordinates 6°27'//N3°23'45'E and covers an area of 999.6km2. Lagos is a municipal area which originated on islands divided by creeks, such as Lagos Island, fringing the southwest mouth of Lagos Lagoon while protected from the Atlantic Ocean by long sand spits such as Bar Beach, which stretch up to 100 kilometres (62 miles) east and west of the mouth. Ikeja is the capital of Lagos State. Traditionally, Lagos State is inhabited by the Aworis subgroup of the Yoruba people but due to high level of commercialization and industrialization, many people from different nationalities work and live within the territory boundary of Lagos State [12]. Lagos is the second fastest growing city in Africa and the seventh fastest growing city in the word. Figure. 1 shows the map of Lagos State showing different locations and the study areas.The stations are IkejaComputer Village, Alaba Market, Westminster (Apapa).

**2.2 Westminster, Apapa.**

Westminster Electronic Market is situated in Apapa Local Government Area of Lagos, Nigeria, close to the Lagos Tincan Island Port (Figure 1). Its location has made it an attractive point for disembarking and selling Used Electrical and Electronic Equipment (UEEE). The market has about 300 outlets where all types of UEEE are sold. Additionally, the market also

**Figure 1** Map of Lagos showing the sample stations

**Figure 2** e-waste at Westminster market dumpsite

has large storing facilities, which make it a big hub for storing UEEE before being redistributed to other markets or exported to neighbouring countries [13] (Figure 2).

### 2.3 Alaba International market

Alaba International market in Ojo, Lagos is one of the known e-waste processing centres in Nigeria. Alaba International market in located in Ojo Local Government Area of Lagos state(Figure 1). Since it is situated in a tropical region, its climate is characterized as warm with plentiful rainfall in season (March-September) and in dry season (October-March). Recycling of e-waste in Alaba International Market in Ojo has taken place for many years, with several tonnes of computer waste handled each year [14](Figure 3).

### 2.4 Ikeja Computer Market

Ikeja Computer Market – popularly known as Ikeja Computer Village located in Ikeja Local Government Area of Lagos State (Figure 1). Aside from the sales of new computer and computer equipment, the sale of used computers, printers and communication equipment is also carried out in the market. Presently, the market is very active in the sales of second-hand products, both in wholesale and retail. The market also deals in the refurbishment and repair operations of computers, phones etc (Figure 4).

### 2.5 Control station (Yaba College of Technology (YCT) Botanical Garden)

Yaba College of Technology Botanical Garden is located in the college premises (Figure 1). It contains well grown plants which are tagged with their scientific names and a green house.The site was chosen because it is devoid of any industrial, electronic waste or facilities that could cause pollution or contamination. The garden is well cared for by staff of the college.

### 2.6 Sample collection

Soil and water samples were collected at different points around the e-waste dump sites in the study areas (Westminster, Alaba International market, and Ikeja Computer Market). Top soil was randomly collected in replicates from each of the three e – waste dumpsites where the unused components are being dumped. The auger was plunged into the ground and the handle turned to collect soil at 0-15cm. The latter step was repeated to collect soil from 15-30cm. Both depths were composited into one sample. The soil samples collected were packed, labeled, preserved and onward transmission to the laboratory for subsequent analysis within 24hours for the following heavy metals Pb, Zn, Cd, Ni, Cr, Cu.

Water samples were collected in replicates from existing wells and boreholes used for drinking and other domestic purposes around the e- waste dump sites. The samples were collected in 500ml sample bottleswhich has been previously washed with detergents and rinsed with distilled water and properly sealed and labelled. Samples were properly labelled and stored in insulated coolers containing ice cubes at 2-6$^0$C and were transferred to the laboratory for analysis within 24 hours. Sampling, preservation and transportation of water samples were carried out under the American Public Health Association (APHA) standard

**Figure 3** *Alaba market dumpsite*

**Figure 4** *Ikeja Computer market dumpsite*

method [15]. The water samples were kept in the refrigerator in the laboratory before heavy metals analysis (Pb, Zn, Cd, Ni, Cr, Cu).

## 2.7 Preparation of the Soil sample for analyses

The soil samples were first air dried for days. The samples were homogenized using a pestle and mortar before sieving through a mesh size of 2mm.Hotplate aqua-regia digestion method was used as described below.

Two grams of each of the homogenized soil samples obtained fromsample preparation procedure abovewere weighed intoa conical flask and 12ml of freshly prepared aquaregia (3ml $HNO_3$ + 9ml HCl) was added. The beakers were covered and the contents were heated for 2 hours on a hot plate. The mixture was allowed to cool and then filtered through a Whatman No. 42 filter paper into a 50ml standard volumetric flask. The filtrate was diluted to 50ml with de-ionized water. The resulting solutions were set aside for heavy metals (Pb, Zn, Cd, Ni, Cr, Cu) analysis using the Atomic Absorption Spectrometry (AAS).

## 2.8 Preparation of the Water sample for analyses

The water samples were shaken to homogenize before sub-sampling for digestion.Twelve ml of conc. $HNO_3$ and HCl were added to 50 ml each of the water samples.The water samples were heated on hot plate to about half the original volume. The flasks were allowed to cool, its contents were filtered into a 50ml standard volumetric each and were made up to mark with de-ionized water. The resulting solutions were set aside for heavy metals (Pb, Zn, Cd, Ni, Cr, Cu) analysis using the Atomic Absorption Spectrometry (AAS) (Perkin Elmer 1100).

*2.8.1 Determination of Heavy metals.* The heavy metals (Pb, Zn, Cd, Ni, Cr, Cu ) in the water and soil samples were determined by Atomic Absorption Spectrophotometer (AAS) method. Samples are presented to the AAS for reading by inserting the Aspirator into the 100mL sample plastic containing the digestate placed on the sample compartment in front of the machine. Upon presentation of samples, the samples get aspirated (sucked) into the Flame through the capillary into nebulizer, gets atomized in the flame, impinged upon by the optimized ray from the Hollow Cathode Lamp (HCL), amplified within the AAS compartment and detected by the Detector within the compartment.The values are read out on the Monitor Display attached to the AAS.

*2.8.2 Quality assurance.*Appropriate quality assurance procedures and precautions were taken to ensure the reliability of the results. Samples were carefully handled to avoid contamination. Appropriate sample preservation/labelling were ensured. Glassware was thoroughly cleaned, and reagents were of analytical grades. Acid digestion was carried out for heavy metal analysis. Deionized water was used throughout the study. Reagent blank determinations were used to correct the instrument readings and repeated calibration of analytical equipment was done.

## 2.9 Statistical analysis

A 3- way between groups ANOVA was used to examine the main effects and interactions of sample locations, sample type (soil and water) and heavy metal types as they relate to concentrations of the heavy metals. Dunnett's multiple comparison test-graph pad was also used.

## 3 RESULTS

### 3.1 Concentration of heavy metals in the water samples.

The concentrations of heavy metals (Pb, Zn, Cd, Ni, Cr, Cu) in the water samples from the various stations; computer village (Ikeja), Alaba International Market and Westminster (Apapa), Control water sample from Yaba College of technology expressed as mg/L are illustrated below(Table 1). The results were compared with the WHO standard for drinking water because the water from the well and borehole sampled are used for drinking purposes.

The concentration of heavy metals (Pb, Zn, Cd, Ni, Cr, Cu) in the water samplesfrom Ikeja, Alaba International Market and Westminster were significantly ($P<0.0001$, $P<0.001$, $P<0.05$) higher than control water samples. Ikeja water samples have the highest concentration of Lead (0.024 mg/L) while Alaba International Market has 0.023 mg/L. Westminster water samples had the lowest lead concentration (0.016 mg/L). The samples from the stations were significantly higher than the WHO (World Health Organization) limit for Pb (0.01 mg/L) and also the Control sample (0.004 mg/L) (Table 1).

Zinc (Zn) concentration was detected in all the water samples including the control (0.016 mg/L). The zinc concentration in Ikeja was 0.111 mg/L while Alaba and Westminster water sample were 0.130 mg/L and 0.154 mg/L respectively. The water samples from the various stations were below the WHO standards (3 mg/L) (Table 1).
Nickel in the water samples was of lowest concentration in the control water sample (0.010 mg/L) while Westminster water samples were found to contain highest concentration (0.026 mg/L) of Nickel. However, Ikeja water samples contain Ni concentration of 0.24 mg/L and Alaba water sample was found to have 0.023 mg/L which was higher than the WHO (World Health Organisaton) limit for Nickel (0.02 mg/L) (Table 1).

Chromium was found in all the water samples for the various stations. However the highest concentration was detected in Ikeja water samples (0.030 mg/L) while the lowest concentration was detected in the control water sample (0.014 mg/L). Chromium detected in water samples from Alaba was 0.019 mg/L and Westminster was 0.023 mg/L which were lower than the WHO limit (0.05 mg/L) (Table 1).

The concentration of copper in the water samples was in lowest concentration in the control water sample (0.016 mg/L) and highest in Ikeja (0.172 mg/L). The concentration of copper in Alaba was 0.154 mg/L and Westminster was 0.100 mg/L. The concentration of copper in all the stations was within the WHO limit for drinking water (2 mg/L) (Table 1). Cadmium was not detected in any of the water samples analysied.

### 3.2 Concentrations of heavy metals in the soil samples

The concentrations of heavy metals (Pb, Zn, Cd, Ni, Cr, Cu ) detected in the soil samples obtained from the various stations; Ikeja, Alaba International Market and Westminster, Control soil sample are shown in Table 2.

**Table 1** *Concentrations of heavy metals in water samples collected from the different stations. Results are expressed as means ± standard deviations (n=5). Statistical significance versus control group:\*\*\* (p< 0.0001), :\*\* (p< 0.01): :\* (p< 0.05).*

| | Water | | | | |
|---|---|---|---|---|---|
| | Control | Alaba | Ikeja | Westminster | WHO limit |
| Pb | 0.004 ± 0.000 | 0.023 ± 0.004*** | 0.024 ± 0.005*** | 0.016 ± 0.005** | 0.01 |
| Zn | 0.016 ± 0.00 | 0.130 ± 0.012*** | 0.111 ± 0.019*** | 0.154 ± 0.011*** | 3 |
| Cd | Not detected | | | | |
| Ni | 0.010 ± 0.00 | 0.023 ± 0.002*** | 0.024 ± 0.004*** | 0.026 ± 0.002*** | 0.02 |
| Cr | 0.014 ± 0.000 | 0.019 ± 0.003* | 0.030 ± 0.003*** | 0.023 ± 0.002*** | 0.05 |
| Cu | 0.016 ± 0.178 | 0.154 ± 0.019 | 0.172 ± 0.019 | 0.100 ±0.000 | 2 |

The heavy metals (Pb, Zn, Cd, Ni, Cr, Cu) concentrations in the soil samples showed a significant difference (P<0.0001, P<0.001, P<0.05) from Ikeja, Alaba International Market and Westminster compared to control soil samples. The Lead (Pb) concentration in the soil samples showed a significant difference (P<0.0001, P<0.001, P<0.05) from Ikeja, Alaba International Market and Westminster compared to control soil samples. Lead (Pb) concentration was highest in Ikeja soil samples (1.360 mg/kg) while the Control soil sample has the lowest concentration (0.005 mg/kg). AlabaInternational Market soil samples contain 1.14 mg/kg of lead and Westminster had 0.98 mg/kg of lead which greatly exceeded theEuropean Union limit (0.3 mg/kg) of heavy metal for lead in soil (Table 2).

Zinc (Zn) concentration was detected in all the analysed soil samples. The control soil samples contain the lowest concentration of zinc (0.005 mg/kg) while the Ikeja soil sample was found to contain the highest concentration of zinc (14.78 mg/kg). The zinc concentration in soil samples from Alaba and Westminster were 9.67 mg/kg and 6.570 mg/kg respectively. The soil samples from the various stations including the control soil sample were above the European Union (EU) limit for soil (0.3 mg/kg) (Table 2).

Cadmium (Cd) was not detected in Westminster soil sample and the Control soil sample. It was found in soil samples from Computer village (Ikeja) and Alaba International Market to be 0.07 mg/kg and9.040 mg/kg respectively which greatly exceeded the European Union (EU) limit for heavy metal concentration of Cadmium (Cd) in soil (Table 2).

Nickel (Ni) concentration was detected in all the analysed soil samples. The control soil sample contain the lowest concentration of Nickel (Ni) (0.35 mg/kg) while the Westminster soil sample was found to contain the highest concentration of Nickel (3.350 mg/kg).

**Table 2** *Concentration of heavy metals in soil samples collected from the different stations Results are expressed as means ± standard deviations (n=5).Statistical significance versus control group:\*\*\*, (p< 0.0001): :\*\* (p< 0.01): :\* (p< 0.05)*

| | SOIL | | | | |
|---|---|---|---|---|---|
| | **Control** | **Alaba** | **Ikeja** | **Westminster** | **EU limit** |
| Pb | 0.005 ± 0.000 | 1.140 ± 0.143*** | 1.360 ± 0.167*** | 0.980 ± 0.126*** | 0.3 |
| Zn | 0.005 ± 0.000 | 9.67 ± 0.689*** | 14.020 ± 4.342*** | 6.570 ± 1.961** | 0.3 |
| Cd | 0.00 ± 0.00 | 9.040 ± 0.042*** | 0.07 ± 0.067* | 0.00 ± 0.00 | 0.003 |
| Ni | 0.350 ± 0.00 | 3.140 ± 1.063*** | 1.560 ± 0.219 | 3.870 ± 1.137*** | 0.075 |
| Cr | 0.450 ± 0.000 | 0.950 ± 0.128*** | 1.390 ± 0.178*** | 0.880 ± 0.115*** | 0.15 |
| Cu | 0.550 ± 0.550 | 5.166 ± 9.134*** | 4.676 ± 6.164*** | 5.779 ± 6.721*** | 0.14 |

Nickel (Ni) concentration in soil samples from Alaba and Ikeja were 3.140 mg/kg and 1.560 mg/kg respectively. The soil samples from the various stations were higher than the European Union (EU) limit for soil (0.75 mg/kg) (Table 2).

Chromium (Cr) concentration was highest in Ikeja soil samples (1.390 mg/kg) while the Control soil sample has the lowest concentration (0.450 mg/kg). Alaba International Market soil sample contain (0.950 mg/kg) concentration of Cr and Westminster has 0.880 mg/kg concentration which greatly exceeded the European Union limit (0.15 mg/kg) of heavy metal for chromium in soil (Table 2).

Copper (Cu) concentration was detected in all the analysed soil samples. The control soil sample contain the lowest concentration of copper (Cu) (0.550 mg/kg) while Westminster soil sample contain the highest concentration of copper (Cu) (5.779 mg/kg). The copper (Cu) concentration in soil samples from Alaba and Ikejawere 5.166 mg/kg and 4.676 mg/kg respectively. The soil samples from the various stations were higher than the European Union (EU) limit (0.14 mg/kg) (Table 2).

There was statistically significant difference between the concentrations of heavy metals in soil and water. In order words, the concentration of heavy metals in soil was statistically significantly higher than that of water (P<0.05).

## 4 DISCUSSION

### 4.1 Metal concentrations and risk assessment

The apparently high concentrations of heavy metals in the soil and water samples from the impacted areas is of particular concern, this is due to the use of the water for domestic purposes by the residents in the area. The concentrations of the heavy metals (Pb, Zn, Cd, Ni, Cr, Cu) in the control sites for both soil and water samples were lower than what was observed in the e wastes dumpsites. This shows that e-waste contribute significantly to the heavy metals concentrations in both water and soil samples.

The level of Pb in the study is high and is a call for concern. Most of the samples from the sites exceeded WHO maximum permissible limits for heavy metals in drinking water. Adverse effects of heavy metals pollution may result from domestic use of the water. High Pb concentration in drinking water may result in metallic poisoning that manifests in symptoms such as tiredness, lassitude, slight abdominal discomfort, irritation and anemia.[16] The results from this study agrees with the findings of Ideriah*et al.*[17] who reported concentrations of Pb between 0.001 and 0.46 mg/L. Lead is found in e-wastes such as in cathode ray tubes, batteries and solders.[12]

Elevated concentrations of Ni in water and soil is due to its presence in electronic wastes in diverse forms, especially in Ni-Cd batteries and electric guns in cathode ray tubes. Although, nickel is an essential element, it is also known to be harmful if inhaled in excess. Nickel at higher concentration is carcinogenic and when inhaled in large amount, it can cause damage to the brain, liver, muscle and kidney.[18] The study agrees with the study by Olafisoye *et al.*[19], who found Ni to range from 0.001 – 1.450 mg/L in water samples investigated. Global input of nickel to the human environment from e-waste sources is mostly from the production of pigments and magnetic tapes.[18]

Zinc (Zn) concentration detected in all the water samples were below the WHO standards (3 mg/L). The soil samples from the various stations including the control soil sample were above the European Union (EU) limit (0.3 mg/L). The high zinc concentration observed in the soil samples at the dumpsites is greatly influenced by the direct release of zinc containing substancesinto the soil. Zinc is a good essential trace element, an excess of zinc intake in the human diet can lead to copper deficiency, immune system disorders.[20] Zn in electronics is found in the interior of cathode ray tube screens as zinc sulphide.[20]

Cadmium (Cd) was below detection limit in Westminster and control soil samples while it was found in soil samples from Computer village (Ikeja) and Alaba International Market to be 0.06 mg/L and 0.04 mg/L respectively, which greatly exceeded the European Union (EU) limit for heavy metal concentration of Cadmium (Cd) in soil. Cadmium (Cd) was not detected in all the water samples investigated. The difference in the cadmium concentrations between the soil and water samples suggest that high cadmium concentration in the soil may be as a result of the direct release into the soil of cadmium from e-wastes containing electrical and electronic equipment such a batteries, PVC, aviation corrosion resistant and light-sensitive resistors. This result is in accordance with earlier studies.[21,22] They reported lowest concentrations of Cd in the samples investigated. This may be due to the ban of Cd-Ni batteries in most countries where electronics are manufactured.[22] The primary health risks of long term exposure are lung cancer and kidney damage.[3]

Chromium was found in all the water samples for the various stations, samples from only one station exceeded the WHO limit for water. Chromium in the soil samples exceeded the European Union limit. Cr and its oxides are widely used because of their high conductivity and anti corrosive properties. Chronic exposure to chromium compounds can cause permanent eye injury.[22]

The concentration of copper in the water samples were lower than the WHO limit of 2 mg/L and copper level in the soil samples from the various stations were higher than the European Union (EU) limit (0.14 mg/L). In a similar study Waalkes *et al.* [23] reported high concentrations of Copper in soil samples investigated.

**4.2 Environmental Forensics assessment**

Further assessment was undertaken in an attempt to establish a pollutant linkage between the e-waste, soils and water samples. This was performed by assessing the average relative proportion of Pb, Zn, Ni, Cr, Cu in the water and soil samples from each site, the results are presented as Figure 5. (Cd was excluded from the assessment as it was only detected in two samples).

These results show that in both soil and water samples the metal signature is different to the three e-waste sites. The signature in the soils and water samples obtained at the e-waste sites are also quite similar indicating that there is a pollutant linkage which is resulting in contamination of the water. However, further research should be undertaken to confirm this linkage.

## 5 CONCLUSIONS

Findings from the study show that the e-waste dumpsites are highly polluted with heavy metals in the following order, Ikeja>Westminister>Alaba. The study also shows that heavy metals pollution of groundwater and soil is an issue of environmental concern, especially when e-wastes are involved. The results revealed the presence of significant concentrations of (Pb, Zn, Cd, Ni, Cr, Cu ) in water and soil in the e-waste dump sites around the three markets investigated when compared to the control sites and WHO standard for drinking water and EU limit for soil. The levels of heavy metals in this study may adversely influence human health since water from the vicinity of the e-waste dumps sites is continuously consumed. The results indicate that the water has been contaminated by the e-waste, however further assessment should be undertaken to confirm the pollutant linkage.

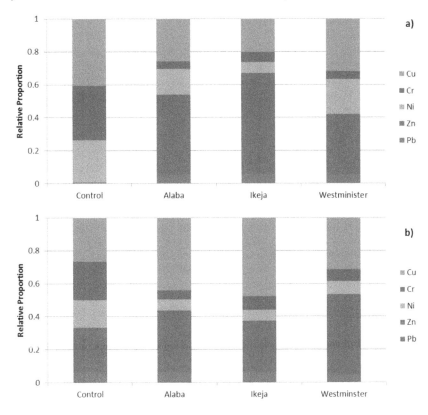

**Figure 5** *Relative proportion of Pb, Zn, Ni, Cr, Cu in a) the water samples and b) the soil samples at the four sites*

## References

1. P. Manomaivibool, T. Lindhqvist and N. Tojo, Extended producer responsibility in a non OECD context: The Management of Waste Electrical and Electronic Equipment in Thailand. *Lund University. Greenpeace International.* 2009, 68 pp
2. . H.-Y. Kang, J.M. Schoenung. Used consumer electronics: a comparative analysis of material recycling technologies. In: *2004 IEEE International Symposium on Electronics and the Environment. Phoenix, AZ, May 10–13, 2004.*
3. I. C. Nnoromand O. Osibanjo. Electronic waste (e-waste): Material flows and management practices in Nigeria. *Waste Management.* 2008, **28**, 1472–1479
4. S. C. Ewuim, C.E. Akunne, M.C. Abajue, E. N. Nwankwo and O. J. Faniran. Challenges of e-waste Pollution to Soil Environments in Nigeria – a review. *Animal Research International.* 2014, **11**(2), 1976 – 1981.
5. L. Anna, W. C. Zong and H.W. Ming, Environmental contamination from electronic waste recycling at Guiyu, south east China. *J Mater Cycles Waste Mang.* 2006, **8**: 21-33.
6. S.E. Musson, Y.C. Jang, T.G. Townsend and I.H. Chung, Characterization of lead leachability from cathode ray tubes using the toxicity characterization leaching procedure. *Environmental Science & Technology.* 2000, 34 4376–4381.
7. J.R. Cui and L.F. Zhang, Metallurgical recovery of metals from electronic waste: a review. *Journal Hazard Mater.* 2008, **158**, 228–256.
8. T. Christine, Recycling Electronic wastes in Nigeria: Putting Environmental and Human Right at Risk. *Northwestern University Journal of international Human Rights.* 2012, **10**(3), 1-11.
9. S. Michael, Nigeria Prominent in Rising Continental E-waste Threat. *Ecojournalism.* 2012, **3**(1), 6-8.
10. A. Manhart, O. Osibanjo, A. Aderinto and S. Prakash, Informal e-waste management in Lagos, Nigeria – socio-economic impacts and feasibility of international recycling co-operations. *The UNEP SBC E-waste Africa Project.* 2011, **3**, 1-127.
11. R. Widmer, O.K. Heidi, S.M. Deepali and B. Heinz, Global perspective on e-waste. *Environmental Impact Assessment.* 2005, **25**, 436.
12. E. A. Ofudje, S. O. Alayande, G. O. Oladipo, O. D. Williams and O. K. Akiode. Heavy Metals Concentration at Electronic-Waste Dismantling Sites and Dumpsites in Lagos, Nigeria. *International Research Journal of Pure and Applied Chemistry.* 2014, **4**(6), 678-690.
13. O.A. Odeyingbo, Assessment of the Flow and Driving Forces of Used Electrical and Electronic Equipment into Nigeria and within Nigeria. Master thesis at BTU Cottbus. Cottbus, 2011.
14. LASEPA, E-Waste Seminar Organized by the Lagos State Environmental Protection Agency titled Electronic Waste and its Effects in the Environment, 7th-8th Nov, 2007.
15. American Public Health Association (APHA). Standard methods for examination of water and wastewater (20th ed.). New York, USA: American Public Health Association 1998.
16. K.M. Cecil, C. J. Brubakar, C.M. Adler, K.N. Diet-Rich, and M. Altaye. Decreased brain volume in adults with childhood lead exposure. *PLoS Med.* 2008, **5**(5), 112

17. T.K.J. Ideriah, F.O. Harry, H.O. Stanley and J.K. Igbara, Heavy metal contamination of soils and vegetation around soild waste dumps in Port Harcourt, Nigeria. *J. Appl. Sci. Environ. Manage.* 2010, **14**(1), 101-109
18. Dater MD, Vashistha RP. Investigation of heavy metals in water and silt sediments of Betwa River. *Interna J Environ Protect.* 1990, **10**, 666-672.
19. O.B. Olafisoye, T. Adefioye and O.A. Osibote. Heavy Metals contamination of water, soil and plants around an electronic waste dumpsite. *Pol. J. Environ. Stud,* 2013, **22**, 5, 1431-1439.
20. B.A. Adelakun and K.D. Abegunde, Heavy metal contamination of soil and groundwater at automobile mechanic village in Ibadan, Nigeria. *Int. J. of Phy. Sci,* 2011, **5**, 1045
21. E.E. Awokunmi, S.S. Asaolu and K.O. Ipinmorati, Effect of leaching of heavy metals concentration of soil in some dumpsites. *Africa. J. Environ. Sci. Tech,* 2010, **4**(8), 495
22. B.A. Adelakan and A.O. Alawode, Contributions of municipal refuse dumps to heavy metals concentrations in soil profiles and groundwater in Ibadan, *Nigeria. J. Of Appl. Biosciences,* 2010, **40**, 2727.
23. M.P. Waalkes, Cadmium carcinogenesis: a review. *Journal of Inorganic Biochemistry*, 2000, **79**, 241.

EVIDENCE FOR ACID MINE DRAINAGE IN AN URBAN STREAM IN THE WEST MIDLANDS

Stephanie M. Turnbull and C.V.A. Duke

School of Biology, Chemistry and Forensic Science, University of Wolverhampton, Wulfruna Street, Wolverhampton, WV1 1LY, UK

1  INTRODUCTION

The purpose of this study was to indicate the type and extent of contamination in Rough Brook stream, Bloxwich, Walsall. The investigation was carried out as Rough Brook appeared to be covered in a thick (approximately 3 mm) iron oxide deposit.[1] The pollution in Rough Brook was thought to be a result of acid mine drainage (AMD), due to the presence of coal mine seams[2] close to the surface in the west midlands. The stream runs directly through a disused landfill which has been developed into a recreational area for the general public. Along the length of the stream are numerous areas of historical industrial works, including an iron and brass foundry[3] and sewage works, which could be accountable for the apparent pollution.

A site investigation was carried out along a selected length of Rough Brook (Figure 1). The site investigation examined contaminant sources and pathways for use in a receptor model[4] for the purpose of identifying potential risks. The study was carried out over a period of six months, consisting of three sampling rounds (11th July 2013, 29th October 2013 and 4th December 2013). Soil and water samples were collected for testing to outline water conductivity, the level of dissolved oxygen, the pH of the water, the pH of the soil and elemental analysis of soil (x-ray fluorescence) and water samples (inductively coupled plasma). Samples were collected over a six month period to take into account seasonal variation due to leaching.[5] Seasonal variation refers to the effect weather conditions have on the concentration of contaminants able to leach from the soil into the stream. With increased rainfall contaminants are likely to become more mobile in the soil resulting in an increase in the concentration of the contaminant recorded in a sample due to an increase in migration.[6] The main objective of this study was to link historical sites of interest to the apparent pollution present in the Rough Brook.

2 EXPERIMENTAL PROCEDURES

**2.1 Sample Collection**

Stream sampling was split into three rounds between July 2013 and December 2013. Water samples collected along the length of Rough Brook were compared to control samples taken from the Wyrely and Essington canal (parallel to Rough Brook on the west side: sample location one) and Ford Brook stream (parallel on the east side: sample locations eight and ten). Control samples were taken to compare and evaluate the results effectively from Rough Brook. A sampling strategy was set out before the site investigation was conducted (Figure 1). The location of each sampling site was monitored and recorded using a handheld global positioning system (Garmin GPS 60) (± 5-8m). Recording the sample location allowed for samples taken during different seasonal rounds to be directly compared with each other; upholding the continuity of the sample collection process.

Sample location six (Figure 1:6-2) was upstream and close to a historical iron and brass foundry (Goscote Works circa 1880). Further sampling sites downstream of Rough Brook 2-2, 3-2, 4-2, 5-2 and 11-2 were located directly against the bank of the disused landfill. Similarly, sample location seven (Figure 1: 7-2) was located next to a sewage treatment outlet.

## 2.2 Soil Samples

Soil samples were collected from the bank of the stream using a metal auger. Approximately 40 grams of soil was collected for subsequent laboratory analysis. A solution of each soil sample was made with distilled water and left to settle for two days.

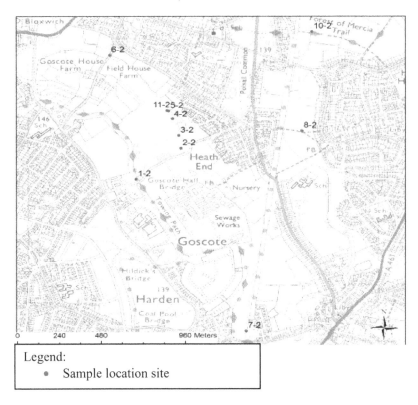

**Figure 1.** *Sampling site GPS locations: indicated on the map via numbered red dots.*

After settling the pH of the water of each sample was recorded using a multi meter (HQ40d Portable, Hach).The soil samples were dried individually in an oven at 40°C for two days before being mixed with Licowax (FluXana) and pressed at ten tonnes of pressure in a die to be analysed by x-ray fluorescence (Xepos, Spectro).

## 2.3 Water Samples

Water samples were collected on site using an extendable pole with a scoop attached. The water samples were transferred into 1 litre (L) plastic containers, where in-situ tests were conducted to outline the dissolved oxygen concentration (%) and temperature (°C) of each sample using a multi meter (HQ40d Portable). Further analysis of the water samples were conducted ex-situ in the laboratory; pH (HQ40d Portable) and elemental analysis via ICP (Ciros, Spectro) were also recorded.

## 3 RESULTS AND DISCUSSION

### 3.1 Soil Results

Individual elemental analysis of the soil samples from all three sampling rounds were analysed via XRF. Using the Tukey HSD comparison test it was outlined that any value of significant difference should be <0.05 at the 95% confidence level. The results concluded that there was no statistical difference (multivariate analysis of variance (MANOVA)) for seasonal variation in soil. MANOVA is a statistical analysis which tests for the difference between the means of data from two or more groups. In this case the three different groups were the sampling rounds (July, October and December).[7] The soil XRF results were collated to note any observable differences between the sampling rounds. It is key to note that the soil samples could not be compared with background samples, as none were collected. Background control samples were unavailable due to access and time constraints related to the project field work. As a result of the absence of background control samples, soil test results could only be compared to those collected in Rough Brook (Table 1, 2 and 3).

**Table 1.** *XRF results obtained from soil samples collected during the July 2013.*

| Sample No. | XRF Soil Mg [ppm] | XRF Soil S [ppm] | XRF Soil Ca [ppm] | XRF Soil Cr [ppm] | XRF Soil Mn [ppm] | XRF Soil Fe [ppm] | XRF Soil Ni [ppm] | XRF Soil Pb [ppm] |
|---|---|---|---|---|---|---|---|---|
| 2 | < 0.031 | 0.4259 | 0.9868 | 0.0193 | 1.214 | 23.28 | 0.1784 | 0.03222 |
| 3 | 0.1396 | 0.8007 | 0.1897 | 0.0131 | 0.0386 | 10.23 | 0.1118 | 0.067 |
| 4 | 0.1869 | 0.4128 | 1.113 | 0.0263 | 0.0571 | 3.972 | 0.01645 | 0.1252 |
| 5 | 0.1546 | 0.4554 | 2.587 | 0.0274 | 0.0951 | 6.485 | 0.02909 | 0.1282 |
| 6 | 0.358 | 0.2014 | 1.785 | 0.0247 | 0.1516 | 2.062 | 0.00491 | 0.00706 |

**Table 2.** *XRF of soil analysis for round one October 2013.*

| Sample No. | XRF Soil Mg [ppm] | XRF Soil S [ppm] | XRF Soil Ca [ppm] | XRF Soil Cr [ppm] | XRF Soil Mn [ppm] | XRF Soil Fe [ppm] | XRF Soil Ni [ppm] | XRF Soil Pb [ppm] |
|---|---|---|---|---|---|---|---|---|
| 2-1 | 0.0818 | 0.465 | 1.326 | 0.01061 | 0.0917 | 10.61 | 0.2511 | 0.04925 |
| 9-1 | n/a | n/a | n/a | n/a | n/a | n/a | n/a | n/a |
| 5-1 | 0.387 | 0.2877 | 1.587 | 0.0129 | 0.1145 | 2.618 | 0.00622 | 0.01533 |
| 6-1 | 0.1967 | 0.4804 | 1.89 | 0.033 | 0.028 | 3.219 | 0.01604 | 0.1141 |
| 4-1 | 0.1564 | 0.3817 | 1.863 | 0.0242 | 0.0848 | 6.991 | 0.02584 | 0.1202 |
| 3-1 | 0.1451 | 0.6041 | 1.343 | 0.1167 | 0.3712 | 9.528 | 0.1382 | 0.1288 |
| 7-1 | 0.1998 | 0.3041 | 1.173 | 0.0124 | 0.1009 | 7.775 | 0.182 | 0.04817 |

**Table 3.** *XRF of soil analysis for round two December 2013.*

| Sample No. | XRF Soil Mg [ppm] | XRF Soil S [ppm] | XRF Soil Ca [ppm] | XRF Soil Cr [ppm] | XRF Soil Mn [ppm] | XRF Soil Fe [ppm] | XRF Soil Ni [ppm] | XRF Soil Pb [ppm] |
|---|---|---|---|---|---|---|---|---|
| 2-2 | 0.1429 | 0.5292 | 1.015 | 0.0196 | 0.0601 | 9.097 | 0.1914 | 0.0789 |
| 4-2 | < 0.39 | 0.3109 | 1.75 | 0.0316 | 0.076 | 7.289 | 0.02494 | 0.1184 |
| 5-2 | 0.319 | 0.3538 | 1.215 | 0.0197 | 0.0797 | 2.627 | 0.00609 | 0.01769 |
| 6-2 | 0.1463 | 0.2373 | 2.309 | 0.0232 | 0.0939 | 8.583 | 0.0304 | 0.0901 |
| 11-2 | < 0.031 | 0.2008 | 1.372 | 0.0709 | 0.2731 | 7.678 | 0.2609 | 0.0478 |
| 3-2 | 0.151 | 0.377 | 1.065 | 0.0839 | 0.0343 | 4.041 | 0.02596 | 0.1419 |
| 7-2 | 0.1977 | 0.4744 | 1.135 | 0.0075 | 0.0802 | 5.689 | 0.0145 | 0.03215 |
| 8-2 | 0.1509 | 0.4553 | 1.253 | 0.038 | 0.0489 | 4.528 | 0.03288 | 0.0917 |
| 10-2 | 0.126 | 0.2008 | 1.435 | 0.1363 | 0.0659 | 6.102 | 0.061 | 0.1434 |

The elemental concentrations in the soil samples were discrete, allowing the trend across Rough Brook to be represented using geographical information systems (GIS) for each element analysed during each sampling round. For nickel, a slightly higher concentration (ppm) was detected downstream, suggesting a possible contribution from the sewage outlet found at sample location 7-2 [8] (Figure 2).

Similarly, the concentration (ppm) of iron was higher in the downstream sample locations for 2-2, 3-2, 4-2 and 7-2 (Figure 1) which were located near coal mine seams in the area. The high concentration of iron could be a result of AMD and the presence of a filled clay pit in the area [9] (Figure 3). The presence of the clay pit was reaffirmed in the copy of an ordnance survey map found dating from 1887 (Figure 4).

The pH results for soil samples ranged from 6.38- 7-64, outlining no indication of a low pH ($<5.0$ )[11] in the stream which could be linked to suspected AMD[12]. Thus demonstrating that significant leaching from the soil surrounding the banks of Rough Brook was not represented in the soil samples collected from the stream.

The results from the XRF analysis outlined that there was a significant statistical difference in the concentration of Ni, Fe and Mg for each of the three (July, October and December) sampling round in the study.

Fe [ppm] XRF for October Results

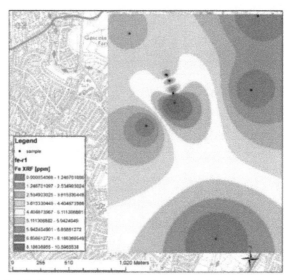

**Figure 2.** GIS of nickel concentrations (ppm) recorded during October 2013 sampling round.

Ni [ppm] XRF for October Results

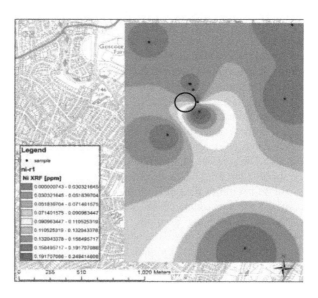

**Figure 3.** GIS of iron concentrations (ppm) recorded during October 2013 sampling round. Black ring identifies the historical presence of a clay pit.

Evidence for Acid Mine Drainage in an Urban Stream in the West Midlands 157

**Figure 4.** *Ordnance survey map from 1887. Rough Brook shown as a dotted black line running from North West to South East. The presence of a clay pit is circled in black.*[16]

## 3.2 Water Results

Statistical analyses were performed on the ICP of water results using multivariate analysis of variance (MANOVA) in statistical package for the social sciences (SPSS); to outline the variance between the sets of results from each round of data collected.

The results illustrated that was a significant difference for all the elements collected in each season analysed. Table 4 portrays the results that under Pillai's trace (the most robust/reliable method) for the seasons were significantly different from each other when taking into account the concentration (ppm) of all the elements recorded for each season. The significance in table 4 below is <0.05.

**Table 4.** *A table to represent the MANOVA results for the ICP of water. Outlining the significance is 0.023 under Pillai's trace.*

**Multivariate Tests**[a]

| Effect | | Value | F | Hypothesis df | Error df | Sig. |
|---|---|---|---|---|---|---|
| Intercept | Pillai's Trace | .993 | 336.496[b] | 8.000 | 18.000 | .000 |
| | Wilks' Lambda | .007 | 336.496[b] | 8.000 | 18.000 | .000 |
| | Hotelling's Trace | 149.554 | 336.496[b] | 8.000 | 18.000 | .000 |
| | Roy's Largest Root | 149.554 | 336.496[b] | 8.000 | 18.000 | .000 |
| Seasons | Pillai's Trace | .964 | 2.211 | 16.000 | 38.000 | .023 |
| | Wilks' Lambda | .036 | 9.636[b] | 16.000 | 36.000 | .000 |
| | Hotelling's Trace | 26.906 | 28.588 | 16.000 | 34.000 | .000 |
| | Roy's Largest Root | 26.906 | 63.903[c] | 8.000 | 19.000 | .000 |

The conductivity of water was generally greater during July, however, stable levels were recorded across Rough Brook. Water conductivity measures the ability of the water sample to pass an electrical current through it. This value is effected by the concentration of dissolved ions available in the water, a decrease in conductivity is related to a period of low water during warmer seasons[13]. The only acceptation to this general trend across Rough Brook was sample location seven (Figure 1: 7-2), where the conductivity was recorded to be higher (Figure 5). This is most likely due to the presence of the sewage outline, resulting in mobile ions in the stream. Figure 5 depicts the conductivity of water results for all sampling rounds and locations. The sample number is listed on the x- axis: 1 represents the canal control sample, 2, 3, 4, 5, 6, 7, 9 and 11 represent the samples collected from Rough Brook and 8 and 10 represent Ford Brook sampling sites. For each sample number there are three vertical bars which represent the three sampling rounds. The first bar representing the July sampling, second bar October sampling and the third bar represents the December sampling.

The pH of the water samples was higher during the July 2013 sampling round, although they were not considered to be acidic. Variation across Rough Brook between sampling locations seems to be minimal as samples two to seven all appear to display similar trends. Ford Brook and the canal control samples provided similar pH values when compared to Rough Brook (Figure 6).

The oxygen concentration % values recorded outline a significant reduction in $O_2$ availability during the October 2013 sampling period. Rough Brook samples contained similar levels of dissolved oxygen relative to the canal control (Figure 1:1-2). However, upstream and downstream variations were noted as evidenced by the downstream decomposition of plant material due to eutrophication [8](Figure 7). It is however, key to note that there was little/no vegetation present upstream for the comparison of downstream vegetation samples.

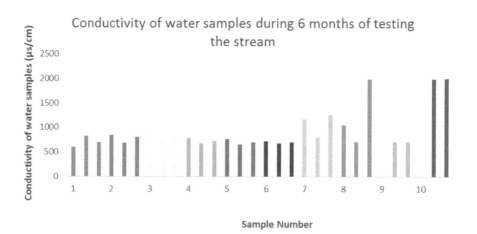

**Figure 5.** *Conductivity of water samples (μs/cm) during six months of testing. For each sample number there are three vertical bars which represent the three sampling rounds. The first bar representing the July sampling, second bar October sampling and the third bar represents the December sampling.*

**Figure 6.** *pH of water samples during six months of testing. For each sample number there are three vertical bars which represent the three sampling rounds. The first bar representing the July sampling, second bar October sampling and the third bar represents the December sampling.*

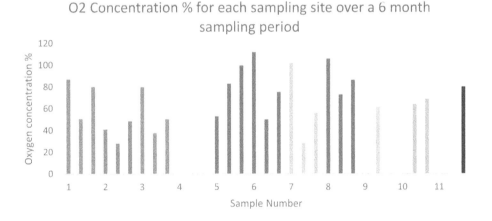

**Figure 7.** *Oxygen concentration (%) of water samples during six months of testing. For each sample number there are three vertical bars which represent the three sampling rounds. The first bar representing the July sampling, second bar October sampling and the third bar represents the December sampling.*

4 CONCLUSIONS

The suspected contamination detected in Rough Brook is likely related to discharges from the iron and brass foundry (Goscote Works circa 1880) and the sewage works (Figure 1). This conclusion was based on the results elemental analysis of water collected in Rough Brook. The concentration (ppm) of Ni, Fe and Mg detected in water samples were statistically different when compared to the control samples taken from the nearby canal and Ford Brook stream, thereby suggesting a relationship between discharges and/or leaching of these elements from soil at the iron and brass foundry into Rough Brook.

Conversely, conductivity values of the water samples in Rough Brook suggested potential contaminate release from the sewage outlet (Figure 1)[14]. Outlining that the results from this study point to two different sources for the contamination present in Rough Brook. The soil results obtained from the study were relatively inconclusive, as we were unable to obtain a control sample for the soils collected off site; the elemental analysis of the soil samples failed to outline any concentrations of significance.

This study provides others with information on how to carry out an effectively simple site review, when investigating the effect historical industrial sites have on the environment.

**Acknowledgments**

I would like to thank the following people for their continued help and support during my undergraduate research: Dr C. V.A Duke, Linda Mc Gowan, Marc Jones, Bart Kluskens, David Luckhurst, Dr Alison Mc Crea, Richard Moss, Diane Spencer, Dr Raul Sutton, David Townrow and Walsall Archives.

**References**

1. Huminicki, D.; Rimstidt, J. D., Iron oxyhydroxide coating of pyrite for acid mine drainage control. *Applied Geochemistry* **2009**, *24* (9), 1626-1634.
2. Sarmiento, A. M.; DelValls, A.; Nieto, J. M.; Salamanca, M. J.; Caraballo, M. A., Toxicity and potential risk assessment of a river polluted by acid mine drainage in the Iberian Pyrite Belt (SW Spain). *Science of the Total Environment* **2011**, *409* (22), 4763-4771.
3. Dungan, R. S.; Dees, N. H., The characterization of total and leachable metals in foundry molding sands. *Journal of environmental management* **2009**, *90* (1), 539-548.
4. Nathanail, J.; Bardos, P.; Nathanail, C. P., *Contaminated land management: ready reference*. EPP Publications: 2011.
5. Driscoll, C. T.; Van Dreason, R., Seasonal and long-term temporal patterns in the chemistry of Adirondack lakes. *Water, Air, and Soil Pollution* **1993**, *67* (3-4), 319-344.
6. Driscoll, C. T.; Lawrence, G. B.; Bulger, A. J.; Butler, T. J.; Cronan, C. S.; Eagar, C.; Lambert, K. F.; Likens, G. E.; Stoddard, J. L.; Weathers, K. C., Acidic Deposition in the Northeastern United States: Sources and Inputs, Ecosystem Effects, and Management Strategies: The effects of acidic deposition in the northeastern United States include the acidification of soil and water, which stresses terrestrial and aquatic biota. *BioScience* **2001**, *51* (3), 180-198.

7. Cosby, B.; Hornberger, G.; Clapp, R.; Ginn, T., A statistical exploration of the relationships of soil moisture characteristics to the physical properties of soils. *Water Resources Research* **1984,** *20* (6), 682-690.
8. Laiho, R., Decomposition in peatlands: reconciling seemingly contrasting results on the impacts of lowered water levels. *Soil Biology and Biochemistry* **2006,** *38* (8), 2011-2024.
9. Lin, C.; Wu, Y.; Lu, W.; Chen, A.; Liu, Y., Water chemistry and ecotoxicity of an acid mine drainage-affected stream in subtropical China during a major flood event. *J Hazard Mater* **2007,** *142* (1-2), 199-207.
10. Walsall Co and Parly. Boro. In *Walsall*, Ordnance Survey, Southampton: Walsall, West Midlands, Great Britain 1922.
11. Helyar, K.; Cregan, P.; Godyn, D., Soil acidity in New-South-Wales-Current pH values and estimates of acidification rates. *Soil Research* **1990,** *28* (4), 523-537.
12. Christensen, J. B.; Jensen, D. L.; Grøn, C.; Filip, Z.; Christensen, T. H., Characterization of the dissolved organic carbon in landfill leachate-polluted groundwater. *Water research* **1998,** *32* (1), 125-135.
13. Daniel, M. H.; Montebelo, A. A.; Bernardes, M. C.; Ometto, J. P.; de Camargo, P. B.; Krusche, A. V.; Ballester, M. V.; Victoria, R. L.; Martinelli, L. A., Effects of urban sewage on dissolved oxygen, dissolved inorganic and organic carbon, and electrical conductivity of small streams along a gradient of urbanization in the Piracicaba river basin. *Water, Air, and Soil Pollution* **2002,** *136* (1-4), 189-206.
14. Thompson, M. Y.; Brandes, D.; Kney, A. D., Using electronic conductivity and hardness data for rapid assessment of stream water quality. *Journal of environmental management* **2012,** *104*, 152-157.

A PRACTICAL EVALUATION OF PETROGENIC AND BIOGENIC METHANE SOURCES IN THE CONTEXT OF REDEVELOPMENT AND EXPLOSIVE HAZARD MANAGEMENT IN LOS ANGELES, CALIFORNIA

Leo M. Rebele and Michael J. Crews

Tetra Tech, Inc., Irvine, California, USA

1 BACKGROUND

In Southern California and in certain regions of the world where methane gas is prevalent due to petrogenic activity, careful planning is needed to manage the potential explosive hazard associated with methane encroachment within confined spaces such as buildings and tunnels. Explosive hazards can occur from either biogenic methane such as associated with historical swamps, or petrogenic methane, often associated with oil fields.[1] From the perspective of mitigation, the type and origin of methane is largely irrelevant. However, from cost allocation perspective, it is often important for a private entity required to install costly mitigation measures as part of their development, to understand the source and type of methane beneath their property. The use of methane forensics can play an important role for property owners and developers in discerning the relative contributions of biogenic and petrogenic methane at a particular property.

2 METHANE MITIGATION REQUIREMENTS – NEW CONSTRUCTION

The City of Los Angeles has specific regulations in place governing new construction within a designated "Methane Zone". These regulations were promulgated in response to a series of events which caused loss-of-life and significant property damage. The most notable examples include the Ross Dress For Less explosion in 1985,[2] which resulted from pockets of methane that had accumulated beneath the building. Another infamous example is the Belmont School incident in 1997, [1,3] where explosive levels of methane were found beneath the school halfway during the construction project, resulting in the most expensive school construction project in history at nearly half a billion dollars. In 2000, a damaging report was published regarding a new residential development in Playa Vista, Los Angeles, which was in the process of being constructed in an area of very high biogenic methane.[4] The findings resulted in large project delays and installations of expensive mitigation systems.

Largely as a result of these highly visible events, the Los Angeles Methane Mitigation Standards were established under Ordinance No. 175790 and Ordinance No. 180619. Specifically, these standards required 1) methane testing using a specified protocol for new construction projects within a designated methane zone, 2) the development of design specifications for mitigation systems and 3) the development of a map of "methane zones". The development of these standards, and their implementation using scientific methods, significantly improved the ability of building officials to pro-actively manage the risks associated with new construction projects within high methane areas.

## 3  METHANE MITIGATION – EXISTING BUILDINGS

As indicated in the previous paragraph, the rules for methane investigation and methane mitigation are relatively straight forward for new construction projects. However, the methane regulations in Los Angeles do not apply to hundreds of thousands of existing buildings in the methane zones, unless a specific hazard (such as direct knowledge of explosive levels of methane) has somehow been identified.

However, many existing buildings (which, as indicated, are exempt from the newer methane building standards), may require special attention when such properties change hands. The fact that the regulation does not in and of itself require investigation or mitigation of methane at a property does not obviate the need for conducting all

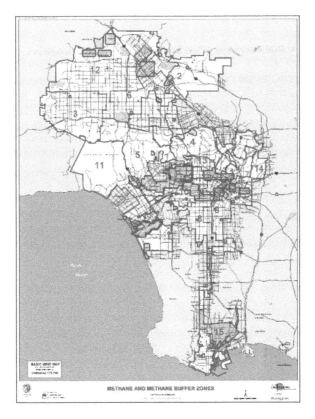

**Figure 1**  *Methane Zones established in the City of Los Angeles illustrating areas of high methane concentrations where special construction measures need to be taken into consideration.*

appropriate inquiry into potential hazards at a Site during a property transaction. New property owners and institutional capital partners look at potential vapour migration and explosive hazards as environmental risks which require mitigation, irrespective of whether such properties are exempt from methane mitigation standards for new construction. It is incumbent on the environmental consultant engaged by a prospective purchaser to identify potential methane hazards during the due diligence period. A subsurface vapour study should be considered if a property with an existing building is identified to be located within the methane zone. If hazardous levels of methane are identified to be present, the property owner can then determine whether mitigation measures are warranted based on 1) their individual risk tolerance and 2) the concentrations of methane found.

## 4 FORENSIC METHODS FOR DESTINGUISHING BETWEEN SOURCES OF METHANE – DEPERSONALIZED CASE STUDY

In cases where the source of methane is clearly from an on-site source, the costs of mitigation measures would generally be expected to be borne by the property owner and/or developer. However, in cases where there is evidence of methane migration from off-site sources, and potential commingling of biogenic and petrogenic sources of methane, the use of environmental forensic techniques and specialized sampling programs can be beneficial in determining responsibility for the mitigation costs.

As an example of demonstrating the usefulness of methane forensics in a redevelopment and transactional context, a depersonalized case study is presented involving three properties (Figure 2). Property A is an active oil and gas producer. Property B is a redevelopment project involving the demolition of a commercial facility and construction of multi-family residential project. Property C is a large multi-family residential project with the intended use of remaining as is. All three properties are located within an LA methane zone, but also happen to be located within the boundary of a historical marsh area. Initial rounds of methane surveys were conducted in conjunction with the redevelopment at Property B in accordance with the LA Methane ordinance, which identified methane concentrations of between 50,000 parts per million by volume (ppmv) and 100,000 ppmv. Concentrations of methane as high as 800,000 ppmv were subsequently detected at the existing apartment complex (Property C) across the street. The developer of the new apartment project and prospective purchaser of the existing apartment complex sought to understand the origin, nature and extent of the methane detected at the properties. Is the methane coming from the historical marshlands as biogenic methane, or was it originating at the oil and gas operation (Property A) in the form of petrogenic methane? Is the oil operator responsible for implementing mitigation measures if the methane was sourced on that property? Are the concentrations stable or increasing or decreasing with time? What kind of mitigation measures could be implemented to protect the existing apartment complex from hazardous levels of methane?

In order to answer these questions, a combination of environmental forensics and scientific sampling methods would be used. The initial phase would consist of collecting methane samples by direct push borings and summa canisters, and submitting to a forensic analytical laboratory for analysis of fixed gas concentrations by ASTM D1946, C1-C5 concentrations by ASTM D1945, and carbon and hydrogen stable isotope ratios by compound specific isotope analysis (CSIA). The forensic analyses would provide answers to the questions about the origin and type of methane. Figure 3 illustrates how isotopic signatures of a methane gas collected from Property A and Property B compares to the signature from a nearby known petrogenic methane gas production well. In this instance,

**Figure 2**  Case study properties located in LA methane zone and historical marsh area.

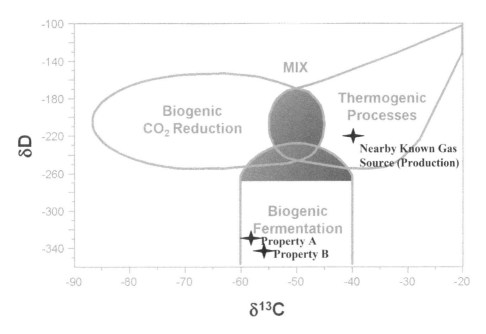

**Figure 3**  Sample data output from CSIA used for differentiating between biogenic and petrogenic methane sources. Carbon and hydrogen isotope ratios of methane in soil gas from a specific site compared to a production gas from a nearby production gas well. Figure after Kaplan et al, 1997 [6]

the graph illustrates how both the methane samples collected from Property A and B are indicative of biogenic methane and does not appear to be associated with the adjacent gas facility (Property C).However, in order to answer the questions regarding how the methane is migrating beneath the properties, monitoring activities can implemented. Given the real estate driven nature of the investigation, it would be important to maintain an element of cost effectiveness to the sampling and monitoring activities. The Gas Clam® (Figure 4) is an instrument that is unique in its ability to collect in situ ground gas measurements over the long term beneath the ground surface.[5] The installation and monitoring of Gas Clam data over a period of several weeks would allow for understanding the subsurface flux (direction and magnitude) of methane concentrations and therefore be able to ascertain the relative threat of the methane in proximity to the existing apartment complex, and the need for methane mitigation measures.

## 5 CONCLUSION

In this depersonalized case study, the results of the methane forensic tests indicated that the methane was indeed biogenic in origin, thus indicating that the oil and gas operation (Property C) was not responsible for the elevated methane detections beneath the residential properties (Property A and Property B). Furthermore, by conducting methane monitoring to observe spatial and temporal variations in the methane concentrations near property lines and building footprints, the prospective purchasers were able to understand the degree to which methane presented a concern to the existing structures, providing the necessary background information to be able to design cost-effective mitigation measures.

The use of environmental forensics and science-based monitoring of methane are important in assisting prospective purchasers and developers of properties within high methane zones understand the environmental hazards associated with the property. Furthermore, possessing a solid understanding of methane conditions and origin help

**Figure 4**  *Image of a Gas Clam® used to monitor temporal variations in subsurface gas concentrations*

owners of real property plan the appropriate mitigation systems for a particular site. This not only ensures that money is not needlessly spent on implementing the wrong mitigation technologies, but also ensures that the property occupants are adequately protected on a long-term basis.

**References**

1   Chilingar, G. V. and B. Endres. Environmental hazards posed by the Los Angeles Basin urban oilfields: an historical perspective of lessons learned. *Environmental Geology*, 2005, **47**(2) 302-317
2   Hamilton, D. H and Richard, L. Cause of the 1985 Ross Store Explosion and Other Gas Ventings, Fairfax District, Los Angeles. *Meehan Engineering Geology Practice in Southern California, Association of Engineering Geologists*, 1992, Special Publication No. 4.
3   Kaplan, I. R. and J. E. Sepich. Geochemical characterization, source, and fate of methane and hydrogen sulfide at the Belmont Learning Center, Los Angeles. *Environmental Geosciences*, 2010, **17**(1), 45–69
4   Brevik, E.C., J.A. Homburg, and C. Tepley.. Stratigraphic reconstruction of the Ballona Lagoon. 1999, Manuscript on file, Statistical Research, Inc., Tucson, AZ, USA
5   Harries, N. 2008. Predicting Ground Gas Risk. The Land Remediation Yearbook. pp33-36.
6   Kaplan, I.R., Galperin, Y., Lu, S., and Lee, R.1997. Forensic Environmental Geochemistry: differentiation of fuel-types, their source and release time. Organic Geochemistry, 27, pp 289-317.

# Author Index

**B**
Balouet, Jean Christophe 31
Barbieri, Cristina 7
Bissolotti, G. 130
Bouchard, Daniel 70
Brosnan, Anne 1
Bruce, John 82
Brutti, R. 130

**C**
Chang, F. C. 22
Cox, Siobhain 92
Crews, Michael, J. 162

**D**
D'Anna, A. 130
Datson, Hugh 82
Doherty, Rory 92
Doherty, V. F. 138
Duke, C. V. A 152

**F**
Famodun, I. O. 138
Fowler, Mike 82

**G**
Gallion, Francis 31
Gigliuto, A. 130

**H**
Hortellan, Marcos Antônio 7
Hsu, Hsin- Lan 22
Hung, H.C. 22
Hunkeler, Daniel 70

**K**
Kwon, Dongwook

**L**
Ladipo, M. K. 138
Liu, P. H. 22

**M**
Martelain, Jacques 31
Martinelli, Luiz 7
McAnallen, L. 92
McIlwaine, R. 92
McLoughlin, Patrick W. 70
Megson, David 31
Morrison, Robert D. 41, 51

**O**
O'Sullivan, Gwen 31

**P**
Pasinetti, E. 130
Peralba, Maria do Carmo Ruaro 7
Peroni, M. 130
Pirkle, Robert J. 70

**R**
Rebele, Leo M. 162

**S**
Sarkis, Jorge Eduardo Souza 7
Scapin, Marcos 7
Smith, Jim 82

**T**
Turnbull, Stephanie M. 152

**V**
Vaccari, R. 130

**W**
Wade, Michael J. 104

SUBJECT INDEX

Acenaphthene, 94, 107, 128
Acenaphthylene, 107, 128
Acid mine drainage, 152
Aerobic, 132, 133, 136-138
Ammonia, 8, 14, 109
Anthracene, 92, 94, 99, 107, 111, 127, 128
Analysis of variance (ANOVA), 94, 95, 144
Anaerobic, 42, 132
Aryl hydrocarbon receptor, 31
Atomic absorption spectrometry (AAS), 9, 12-15, 144

Bacteria, 58, 132, 135-138
Benzene, 62, 71, 75, 92, 132
Benzo(a)pyrene, 8, 15, 17, 112, 115, 116, 118, 120, 127, 128
Benzo(a)anthracene, 112, 115, 116, 118, 120
Benzo(b)fluoranthene, 127, 128
Benzo(e)pyrene, 107
Benzo(k)fluoranthene, 15, 17, 128
Benzo(g,h,i,)perylene 15, 17, 128
Biodegradation, 133, 134, 136, 137

Cadmium, 139, 145, 146, 148
Carbon tetrachloride, 44
Carcinogen, 8, 51, 92, 148
Chlorinated solvent, 22, 41-47, 51, 55, 63-65, 71
Chromium, 9, 42, 139, 145, 147, 148, 149
Chrysene, 15, 17, 95-97, 107
Coal tar, 104-112, 118, 121, 125,
Compound specific isotope analysis (CSIA), 22, 24, 41, 44, 58, 65, 70, 78-80, 164, 165
Comprehensive two dimensional chromatography (GCxGC), 34, 35, 102
Copper, 9, 12, 45, 145-149

Dense non-aqueous phase liquid (DNAPL), 22-30, 56-58
Dendrochemical, 37
Dibenzo(a,h)anthracene, 128
Dichlorethane (DCA), 29, 41-46, 75
Dioxin, 31, 32, 35, 62, 63, 140
Dissolved inorganic carbon (DIC), 8

e-waste, 139-142, 148-150
Electrolytic conductivity detector (ELCD), 25, 28
Electron capture detector (ECD), 25, 28, 29, 34
Expert witness, 1, 4, 5
Exploratory data analysis, 34, 52, 66

Fertiliser, 1
Flame ionisation detection (FID), 106
Flame retardants, 139
Fluoranthene, 15, 17, 95, 101, 107, 110-112, 115-120, 125-128
Fluorene, 107, 128
Furan, 31, 32, 62
Fugacity, 92-101

Gas Chromatography (GC), 4, 25, 34, 35, 73, 74, 94, 102, 106, 108
Groundwater, 8, 19, 22-28, 51, 52, 57-61, 65, 70, 71, 98, 110, 111, 118, 131, 132, 137, 150

Hexachloroethane, 42, 44
Hexachlorobenzene, 44,
Human health, 1, 2
Hydrocarbon, 4, 11, 15, 22, 32, 42, 45, 46, 53, 60, 92, 93, 104, 105, 111, 115, 131-138

Indeno(1,2,3-c,d)pyrene, 15,17, 92, 95-99, 107, 128
Inductively coupled plasma (ICP), 15, 84, 85, 152, 154, 157,

Landfill, 7-15, 16, 18, 19, 152, 153
Leachate, 7-16
Lead, 9, 139, 145-148
Legislation, 3, 7, 8,

Mass spectrometry (MS), 15, 25, 34, 35, 73, 74, 83-85, 93, 106,
Metal, 7-20, 22, 23, 42, 58, 59, 71, 131, 139-150, 153
Methane, 41, 162-166
Micro-bubble, 131

Microbiological, 132, 136, 138
Microscopy, 64, 84
Multiple linear regression, 88, 104,
Multivariate analysis of variance (MANOVA), 154, 157

Naphthalene, 15, 17, 94, 99, 107, 108-128
Nickel, 145-148, 155 156
Non-aqueous phase liquid (NAPL), 26, 70, 72, 76, 118-120

Partitioning coefficient, 71, 75
Photoionisation detector (PID), 29, 73
Phenanthrene, 15, 17, 92, 95, 99, 107, 111-114, 128
Pesticide, 1, 4, 42, 58
Petroleum, 1, 53, 55, 62-64, 92, 99, 105, 108-111, 115-128, 131, 137
pH, 133, 152-155, 158, 159
Polychlorinated biphenyls (PCBs), 31-39, 62-64,
Polycyclic aromatic hydrocarbon (PAHs), 8-20 92-101 104-128
Principal component analysis (PCA), 12-20, 34-36, 66, 104, 111-119, 123-127
Pyrene, 15, 17, 92-101, 107, 110-120, 125, 128

QEMSCAN, 84-86

Sediment, 7-20, 61, 93, 104, 110, 115-124, 128
Sewage, 9, 14-18, 56, 152-155, 158, 160
Semi-volatile organic compounds (SVOCs), 92, 94, 102
Solid phase microextraction (SPME), 25
Stable isotope, 7, 19, 20, 25, 58, 62, 65, 164,
Surface water, 4, 8

Tributyltin (TBT), 1
Trichloroethane (TCA), 42, 43, 47, 61,
Trichlorethylene (TCE), 22, 41, 44, 45
Tetrachlororethane (TeCA), 41-47
Tetrachloroethylene (PCE), 22, 41
Toxic equivalency factor (TEF), 31, 32, 39

Vinyl chloride, 42, 44, 46, 56, 136
Volatile organic compound (VOC), 25, 70-79

Waste Electrical and Electronic Equipment (WEEE), 139,

X-ray Fluorescence (XRF) 11, 152, 154

Zinc, 9, 45, 145-148